Emerging World

The Evolution of Consciousness and the Future of Humanity

Roger P. Briggs

EMERGING WORLD: THE EVOLUTION OF CONSCIOUSNESS AND THE FUTURE OF HUMANITY

Copyright © 2021 Roger P. Briggs

COVER ART: "Earth and Sky" by Emily Silver

1405 SW 6th Avenue • Ocala, Florida 34471 • Phone 352-622-1825 • Fax 352-622-1875
Website: www.atlantic-pub.com • Email: sales@atlantic-pub.com
SAN Number: 268-1250

No part of this publication may be reproduced, stored in a retrieval system, or transmitted in any form or by any means, electronic, mechanical, photocopying, recording, scanning, or otherwise, except as permitted under Section 107 or 108 of the 1976 United States Copyright Act, without the prior written permission of the Publisher. Requests to the Publisher for permission should be sent to Atlantic Publishing Group, Inc., 1405 SW 6th Avenue, Ocala, Florida 34471.

Library of Congress Control Number: 2020923340

LIMIT OF LIABILITY/DISCLAIMER OF WARRANTY: The publisher and the author make no representations or warranties with respect to the accuracy or completeness of the contents of this work and specifically disclaim all warranties, including without limitation warranties of fitness for a particular purpose. No warranty may be created or extended by sales or promotional materials. The advice and strategies contained herein may not be suitable for every situation. This work is sold with the understanding that the publisher is not engaged in rendering legal, accounting, or other professional services. If professional assistance is required, the services of a competent professional should be sought. Neither the publisher nor the author shall be liable for damages arising herefrom. The fact that an organization or website is referred to in this work as a citation and/or a potential source of further information does not mean that the author or the publisher endorses the information the organization or website may provide or recommendations it may make. Further, readers should be aware that Internet websites listed in this work may have changed or disappeared between when this work was written and when it is read.

TRADEMARK DISCLAIMER: All trademarks, trade names, or logos mentioned or used are the property of their respective owners and are used only to directly describe the products being provided. Every effort has been made to properly capitalize, punctuate, identify, and attribute trademarks and trade names to their respective owners, including the use of ® and ™ wherever possible and practical. Atlantic Publishing Group, Inc. is not a partner, affiliate, or licensee with the holders of said trademarks.

Printed in the United States

PROJECT MANAGER: Kassandra White
INTERIOR LAYOUT AND JACKET DESIGN: Nicole Sturk

Contents

Begin Here 1

PART I – THE STORY OF HUMANS

CHAPTER 1: The Science of Evolution 7
CHAPTER 2: Consciousness 101 53
CHAPTER 3: Structures of Consciousness..89
CHAPTER 4: The Macro-stages..117

PART II – THE FUTURE OF HUMANITY

CHAPTER 5: The End of an Era 159
CHAPTER 6: Seven Markers 187
CHAPTER 7: Emerging World 227

Appendices..277
Index..293
Acknowledgements...301
About the Author...303

Contents Details

Begin Here 1

PART I – THE STORY OF HUMANS

CHAPTER 1: The Science of Evolution 7
   The Evolution of Life..7
      The WHAT and WHEN of Evolution... 8
      The HOW and WHY of Evolution 10
      Controversy and Debate... 13
      Creative Emergence 17
   The Evolution of the Universe.. 19
      How Old is the Universe?.. 20
      What is the Universe Doing? 22
      What are Stars? 24
      The Big Picture 28
   The Evolution of Culture.. 31
      The Story of Genus **Homo** 35
      The Great Transformations.. 47
      Culture and Consciousness 49

CHAPTER 2: Consciousness 101 53
   Consciousness in Philosophy 54
   Consciousness in Psychology 58
   Consciousness in Neuroscience.. 62
   Consciousness in Physics 68
   Consciousness in Buddhism... 75
   Consciousness Outside the Brain 81
   Conclusion 86

CHAPTER 3: Structures of Consciousness..89
   The Ever-Present Origin: Jean Gebser 92
   The Five Structures 94

The Archaic Structure 99
The Magical Structure 100
The Mythical Structure 102
The Mental Structure 106
The Integral Structure 111
Stepping Back from the Five Structures 115

CHAPTER 4: The Macro-stages 117
Origins of the Modern Mind: Merlin Donald 117
Stages of Human Evolution 123
The Reflexive Universe: Arthur M. Young 131
Reflexive Evolution: The Seven Stages of Process 132
The Seven Kingdoms of Nature 134
Fractal Evolution: The Sub-stages 141
Applications of the Theory of Process 146
Summary of the Theory of Process (for our purposes) 147
Grand Synthesis: *The Gebser-Donald-Young Macro-stages of Human Evolution* 148

PART II – THE FUTURE OF HUMANITY

CHAPTER 5: The End of an Era 159
Owning the Material Stage 161
The Story of Physics 171
The Twentieth Century 178
The Shift from Material to Planetary 180
A Speculation: Life in the Universe and the Technological Bottleneck 182

CHAPTER 6: Seven Markers 187
1. Awakening 189
2. Connectedness 192
3. Meta-perspective 200
4. Softening of the Ego 203
5. Wholeness 211

6. Heart Opening 215
7. The Evolutionary Worldview 222

CHAPTER 7: Emerging World 227
    20th Century Economics 228
    Doughnut Economics 230
    The Ecological Civilization 238
    The Club of Rome 241
    The Well-being Economy Alliance 243
    The Earth Charter 244
    From *What* to *How* *246*
    The Culture Project 248
    Healing from Materialism 250
    Planetary Spirituality 258
    The Planetary Human 263
    The Power of Personal Choice 265
    Large-scale Initiatives for Cultural Transformation 269
    Epilogue 274

Appendices 277
    Appendix I – Suggested Readings 277
    Appendix II – Key Descriptors Summarizing Gebser's Structures and Stages 280
    Appendix III – Diving Deeper into Arthur Young: The Torus and Seven-ness 280
    Appendix IV – Summary of the Gebser-Donald-Young Macro-stages 282
    Appendix V – The Earth Charter 283

Index 293

Acknowledgements 301

About the Author 303

Begin Here

In 1951 when I was born, the future of humanity looked very bright. Following two World Wars and the Great Depression, the '50s and '60s brought hard-earned prosperity to the U.S. in the form of big cars that guzzled 25-cent-a-gallon gasoline and sprawling homes in the suburbs tended by June Cleaver; "better living through chemistry" and a television in every living room was the essence of the good life; by the end of the '50s the space age had begun, and the sky was literally no longer the limit. In those days everyone thought the progress possible for humans was boundless, and the seemingly unlimited resources of our planet were ours for the taking. It was a small world on a big planet.

But today, the future of humans on this planet looks very different. It's now a *big world on a small planet*.[1] For the first time, humans recognize the very real possibility that our future could be very bleak and fraught with suffering, or we may have no future at all. The pandemic of 2020 and its devastating repercussions exposed many deficiencies in our civilizational infrastructures and flaws in our cultural paradigms; it is now clear to all whose eyes are open that our current path is not sustainable, and it's time for a new way for humans to be on this planet, and a new relationship *with* this planet.

---

1. Taken from the excellent book by Johan Rockström and Mattias Klum, *Big World, Small Planet*. Max Strom Publishing (2015).

But this is not a new idea. In 1966 when I was fifteen, there was something in the air, and many of us became infected with a dream and a shared vision of a new world. This became known as the counterculture. There was much talk and many great songs written about love and peace and justice, and some real changes like the Civil Rights movement, the women's movement, and environmentalism. Yet today war, poverty, and injustice are still here, and our planetary support systems are collapsing. Many now say that these dreams of a new world were simply naïve because human nature will never really change; and in fact, this dream *did* fade away for most of my generation, a youthful fantasy that evaporated with the realities of adulthood and the seduction of materialism.

But it did not fade for me. More than fifty years later some of us still carry that vision, and it is reawakening today in a new generation of dreamers who see clearly that things cannot continue as they have been. Across all generations and all people, it is now time to come together to complete the work that began in the '60s, to co-create a new civilization that works for everyone and honors our living planet. *Emerging World* is a contribution to this effort and is intended to inspire hope for the future, especially for the young ones of our planetary tribe who will inherit this world.

Since I was a teenager, I have been fascinated with the phenomenon of *humanity* – how we came to be on this miraculous planet, what makes us different from other animals, and what our future could hold. But as a lover of science, and having great curiosity, I could not be satisfied with simply accepting that God is the all-encompassing answer; and as an intuitive person, I could not accept that we are here merely by random chance. So I began a journey and a search that became this book.

Emerging World is structured in two parts, around two big, fundamental questions:

> *What are humans and what is our story?*
> *What is the future of humanity?*

Part I, *The Story of Humans*, begins by introducing the two major concepts that support the rest of the book: *evolution* (Chapter One) and *consciousness* (Chapter Two). Thereafter, these merge into the main subject of the book, the *evolution of consciousness*.

In Chapters Three and Four we explore the pioneering work of Jean Gebser, Merlin Donald, and Arthur M. Young, who each contributed groundbreaking theories about the evolution of culture and consciousness, and the story of humans. By the end of Chapter Four, the end of Part I, these three theories are integrated and correlated with the archaeological record to create a new and simple four-part story of human evolution – *our story*. This story begins roughly 3 million years ago when our lineage diverged from the rest of life on the planet in a new direction, to go where no other living thing on Earth has ever gone. Our ancestral lineage, which I will refer to by the scientific name *Homo*, became something entirely new. As a result, we humans are now an overwhelmingly powerful force on this planet. We dominate. And in one way or another we will determine our own future, whether that be a bad one or a good one.

In Part II, *The Future of Humanity*, we begin by confronting head-on our current civilizational paradigm and recognizing it as a developmental stage that we have been in for 5,000 years. We then explore the emerging higher consciousness and the next stage of human evolution that many visionaries have foretold. We will refer to this as the *Planetary consciousness*. The last chapter takes on the project of our time: building a new civilization that promotes the well-being of all people and the planet by generating a *Planetary culture*. In the end, I hope to leave you inspired and optimistic about the future of humanity, and committed to actions you can take that support and nurture this emerging world.

Before we begin this journey together, I want to offer some words of encouragement for those readers who might feel intimidated or discouraged by the amount and the nature of knowledge I will take you through, especially in the first four chapters. This brings to mind a question I heard many times during my 30-year career as a physics teacher. *Will this be on the test?* For too many people the experience of learning in school was fraught with stress and fear and anxiety when, instead, learning and the exploration of knowledge should be joyful and liberating and exciting. So, I promise you there will be no test, and I ask you to try to feel the joy and don't let anything intimidate you as we explore these vast fields of knowledge together. Simply, let it wash over you. It is certainly not the case that you will need to remember and master every detail before being able to move on. Far from it: Mostly I will be fleshing out big ideas, with some supporting details, trying to bring you the forest rather than the trees. If there are any important points that you will need to remember for later, I will emphasize those. This will be a wild ride, so fasten your seat belts, and let our journey begin!

Roger Briggs
Boulder Canyon, Colorado
January 1, 2021

Part I

The Story of Humans

CHAPTER 1

The Science of Evolution

The Evolution of Life

No one knows how life got started on this planet or if this has happened anywhere else in the universe. Somehow, organic molecules deep in the oceans of an infant Earth organized into microscopic cells with a permeable membrane that separated the inside environment from the outside environment, along with a power system and the ability to self-replicate relentlessly. How each of these components could emerge, while working together to make a living organism, remains one of the greatest mysteries in science. We don't even know what life actually is. It is clearly made up of matter — atoms and molecules — but it is quite different from other forms of matter, like rocks, rivers, and air. The form of matter we call life is animated with *something* that makes it grow and respond to its environment and reproduce and *evolve* over time.

As much as we don't know about life, there are quite a few things we *do* know about it. The earliest known evidence of life comes from chemical signatures in ancient rocks that are about 3.8 billion years old, and direct evidence of life — fossil bacteria — can be found in rocks that are about 3.5 billion years old. We know that life took hold on Earth sometime between about 3.5 and 4.0 billion years ago.

But the magnificent plants and animals we see everywhere today are nothing like bacteria. Early life didn't just stay the same — it got much more complex and much larger. Somehow, over billions of years, bacteria gradually turned into conscious beings, who write symphonies and can travel to the Moon. This is the evolution of life.

The WHAT and WHEN of Evolution

Paleontologists have constructed a very detailed account of the evolution of life based on fossils of once-living things (1-1) that have been preserved in rock for millions or even billions of years. This is known as the fossil record. Fossils from all over the world fit together remarkably well to give us a coherent and consistent picture of what life looked like over billions of years as it continually evolved. There are also many gaps in the fossil record, but the big picture is very clear. This is summarized below in 1-2 and should look familiar if you took a biology class. This is the *what* and the *when* of evolution — what happened and when it happened — and it's about as certain as the Earth is round.

1-1
A Trilobite fossil, 514 million years old, found recently in China.
Credit: Hopkins, MJ, et al. PLoS ONE 12(9): e0184982.

It's difficult to comprehend billions of years, so the right-hand column in Table 1-2 puts the chronology of life into more understandable terms. If we think of the entire lifetime of Earth (4.6 billion years) as a single 24-hour day, we can give each event a clock time. You can see that the evolution of life proceeded slowly for the first few billion years but has been accelerating dramatically over the last half-billion years. The action really picks up after about 9:00 in the evening with the so-called *Cambrian Explosion*, when life began to look like the things we see today, and the emergence of human beings (*Homo sapiens*) about 6 seconds before midnight took evolution in a whole new direction, at an ever-faster pace. Even more amazing is that all of "history" as we learned it in school — from the first civilizations in Mesopotamia, to the Babylonians, Egyptians, Greeks, Romans, and so on — fills only the last one-tenth of a second before midnight!

1-2

HIGHLIGHTS IN THE EVOLUTION OF LIFE

Event	How Long Ago bya = billion years ago mya = million years ago	"Earth Day" Time If the lifetime of our planet was a single 24-hour day
A. Earth forms	4.6 bya	12:00 am (midnight)
B. First oceans	3.9 bya	4:15 am
C. Earliest evidence (chemical) of life	3.8 bya	4:45 am
D. Earliest known fossil bacteria	3.5 bya	5:30 am
E. First complex (Eukaryotic) cells	2.4-2.6 bya	11:00 - noon
F. First multi-cellular life	1.9-1.5 bya	2:00 — 4:00 pm
G. Plants, animals, and fungi diverge	1.5 — 1.2 bya	4:00 — 5:40 pm
H. *Cambrian Explosion* — large life forms appear	540-510 mya	9:10 — 9:20 pm
I. First vertebrates (fishes)	510 mya	9:20 pm
J. First life on land (plants)	475 mya	9:30 pm
K. First animals on land (bugs)	415 mya	9:50 pm
L. First Amphibians	350 mya	10:12 pm
M. First Reptiles	320 mya	10:18 pm
N. First Mammals	248 mya	10:42 pm
O. First Dinosaurs	230 mya	10:48 pm
P. First Birds	160 mya	11:10 pm
Q. Dinosaur extinction (asteroid strikes Earth)	65 mya	11:39 pm
R. Mammalian Explosion	60 mya	11:41 pm
S. First Primates	55 mya	11:43 pm
T. First Apes	20 mya	11:54 pm
U. First Gorillas	9 mya	11:57 pm
V. *Homo* lineage begins (first tool-making)	3 mya	11:59 pm
W. First *Homo sapiens* (<u>anatomically</u> modern humans)	300,000 ya	11:59:54 pm
X. First <u>behaviorally</u> modern humans	60,000 ya	11:59:59 pm
Y. First civilizations — traditional history begins	5,200 ya	11:59:59.9 pm
Z. Now	Zero ya	12:00:00 am (midnight)

Evolutionary biologists have organized this chronology into a tree of relationships among all living things — a simplified version is shown below. In this tree, life began at the trunk of the tree then proceeded upward, splitting into the many branches containing all of the life forms we know of.

1-3

A HIGHLY SIMPLIFIED EVOLUTIONARY TREE OF LIFE

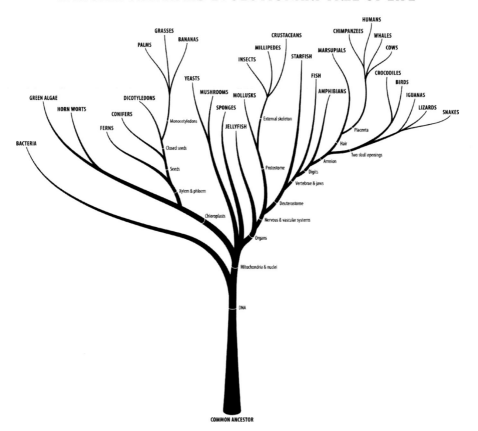

The HOW and WHY of Evolution

How did primitive bacteria turn into you and me? This is a remarkable feat, even if it did take almost 4 billion years. Science does have an explanation in the form of Darwin's theory of evolution. Though Charles Darwin is credited with the theory, Alfred Wallace worked in parallel with Darwin —

the two corresponded — but Darwin's *On the Origin of Species* (1859) took the world by storm. It was controversial from the beginning, and it still is today, but within the majority of the scientific community, it is taken as the foundation of evolutionary biology. Many of us learned it in school. Darwin's ideas have been further supported by the discovery of DNA in 1951 and the exploding field of molecular genetics.

Darwin's key insight was that what drove evolution forward was the 1-2 punch of *descent with modification,* followed by *natural selection.* Darwin observed that offspring are always slightly different from their parents — this is what descent with modification means. He did not know what caused this, but today, knowing about DNA, scientists understand that the information encoded in the DNA molecule gets slightly altered as it is passed from parents to offspring by things like cosmic rays from outer space, copy errors, local sources of radiation, and other disruptions from the environment. These are *random* changes in the parents' DNA, known as *mutations.* The genes of parents can also be altered by various *recombination* processes during reproduction, in which sections of DNA are shuffled and exchanged across paired chromosomes. The random changes in the genome caused by both mutation and recombination create the variation and novelty that evolution requires. Nature creates many variations through these random processes, and they are thrown out to the environment, the harsh *natural* world, where they either succeed or fail.

This sets up the second part of Darwin's mechanism, *natural selection.* Nature (the environment) "selects" the best offspring and eliminates the worst according to which ones *survive to reproduce.* This is commonly known as "survival of the fittest," a phrase coined by Herbert Spencer (not Darwin). Offspring that survive and reproduce will pass along their "good" genes, but if an organism can't survive to reproduce, its genes are eliminated. Nature selects organisms that survive to reproduce. Survival is all that matters. At least that's how it was until humans came along.

Natural selection works most effectively in harsh, challenging environments where many organisms can't survive. Only the very best do survive, and best means *best adapted to their environment.* For example, suppose a particular plant species is thriving in a certain location, but then a severe drought sets in for a few years, so that most of the plants die. But a few survive for some reason — a mutation of some sort has taken place that makes them more drought resistant — and because they have survived, these plants can reproduce and pass along their drought-resistant genes to their offspring. Their offspring will also be drought resistant, and this beneficial mutation will be passed along to future generations. Of course, the plants that died before they could reproduce could not pass along their genes. Nature eliminated them from the gene pool and selected the plants that were best able to handle the environment. These slightly different plants that survived have adapted to a new environment. They have evolved. Darwinian evolution proceeds one step like this at a time, each step the result of a random mutation.

Many mutations are useless (or neutral), and a few are detrimental to survival, but the occasional mutation that enables survival in a changing environment is what drives evolution forward. In tiny steps, each one the result of a small random change in the genome, the evolution of life inches forward.

This mechanism of small random changes in the genome acted on by natural selection is made all the more powerful by two other things: numbers and time. Most organisms make huge numbers of offspring — think of the seeds of a dandelion scattered by the wind. The more possibilities thrown out to the environment, the greater the chances that one of them will enhance survival. The second is time: Life has been doing this for almost 4 billion years. That is a *long, long* time, and presumably long enough for tiny random changes acted on by natural selection to accumulate and gradually turn bacteria into humans.

We know that Darwin's two-part mechanism does actually work, and we've seen it in action many times. For example, when penicillin became widely available in 1941, it worked very effectively to kill *Staphylococcus* bacteria, the cause of deadly staph infections. But it never kills every single bacterium because a few mutate in some way that allows them to survive. When they reproduce, their offspring will also be resistant to penicillin. This is an ongoing problem with antibiotic medicines today.

The modern synthesis of Darwin's two-part mechanism, along with molecular genetics, is the core explanatory model in science today for the evolution of life. But increasingly, some scientists are questioning whether this is a *complete* explanation. Could this mechanism *alone* turn bacteria into humans? Because it took place in the past, over billions of years, we can't test the theory, nor can we repeat the experiment. There are still baffling things about the evolution of life that are now causing many scientists to look deeper for a more complete explanation.

Controversy and Debate

The modern scientific theory of evolution described above is highly controversial in some parts of the U.S. (but hardly anywhere else), where religious groups have tried to ban its teaching in schools. What's the problem? At one extreme is the "young Earth" view[2] that says God created Earth and all its inhabitants some six thousand years ago (4004 BCE[3] to be exact). But we have overwhelming evidence that life has been around for almost 4 billion years, and the fossil record shows clearly that it *did* evolve over this time into more and more complex forms, as detailed in Table 1-2. We can, therefore, dismiss the young Earth view as impossible.

However, there is a growing debate within the scientific community about whether evolution is strictly a random walk with no direction or purpose and is, therefore, unpredictable or that it tends toward certain outcomes, as though with purpose. This first position, the random unpredictable walk, is called *contingent evolution*, meaning evolution by chance. The opposite position, known as *convergent evolution*, holds that evolution tends to *converge* at certain outcomes, as though pulled or pushed in some way.

---

2. A 2017 Gallup Poll found 38% of Americans holding the young Earth view. But this was the *lowest* number in 35 years of polling this question.

3. This figure originated from Archbishop James Ussher (1581-1656) of the Church of Ireland who derived it from the genealogies in the book of Genesis. The figure made it into the King James Bible and has stuck around ever since.

The late evolutionary biologist Stephen J. Gould proposed a thought experiment that captures this debate in his best-seller from 1990, *Wonderful Life*. He asked, if you could "rewind the tape" of evolution, start over from any point in the past, and let evolution play out naturally from there, would things turn out about the same as today or very differently?

For Gould, the answer was clear and simple: things would turn out very differently every time you replayed the tape of evolution. It would be *very unlikely* that we would ever end up twice with humans or anything that looks much like today's life. Gould felt that evolution is contingent — it is a long string of chance events with no direction, no mandate, and no purpose except to promote survival. A few decades ago, this was the majority view within biology.

However, those who hold the view of *convergent* evolution would give the opposite answer to Gould's question: Things would come out very similarly every time the tape of evolution was replayed. This view is championed by evolutionary biologist Simon Conway Morse in his 2003 book, *Life's Solution — Inevitable Humans in a Lonely Universe*. As the title suggests, Morse asserts that evolution would produce human-like creatures every time the tape was replayed. It is in stark contrast to Gould's view of contingent evolution, where humans are considered a one-time result. These two biologists had lively debates over the question of contingent versus convergent evolution, but, largely, the mainstream science community dismissed Morse and agreed with Gould.

Today this debate has been rekindled with new evidence that convergent evolution is real, and some troublesome questions still linger. Consider, for example, the human eye. Darwin himself was troubled by the complex eye (as we have) and how it could have evolved in tiny steps through his mechanism. Like us, many animals have a complex eye, consisting of a lens, a cornea, a pupil, and a retina that all work together. This combination focuses light precisely on the retina, which is packed with photoreceptors that generate voltage pulses that travel to the visual cortex of the brain, where the experience of vision is then created (somehow). How could all these mechanisms evolve separately in small steps, where each step promoted survival, and they end up working together so flawlessly?

Even more puzzling is that the complex eye evolved many different times in *completely separate lineages*, such as squids, snails, jellyfish, spiders, and vertebrates. It's a challenge to explain how it could happen once, and seemingly impossible to explain how it could happen many different times if evolution is indeed a random walk. The evolution of a complex eye independently in each of these lineages is like a separate replaying of the tape; clearly things came out very similarly in each replay. Evolution converged at the same mechanism for sophisticated vision many times. How does this happen? We don't know.

When Gould posed his famous thought experiment in 1990, he also lamented that the experiment couldn't possibly be performed — we can't actually replay evolution. But since then, many experiments *have* been done that represent a replaying of the evolutionary tape. Organisms with fast generation times, like *E. Coli* bacteria, are ideal for running parallel evolution experiments. Researchers start with a number of identical populations of *E. Coli* cloned from the same parent, each in its own closed environment. Then they turn evolution loose; they let each population evolve over time and see what happens. The generation time for *E. Coli* is about 3.6 hours, so in a reasonable amount of time evolution can run for many generations. One of the earliest experiments of this type started in 1988 with 12 parallel populations of *E. Coli* and has been running ever since, now having evolved through 65,000 generations[4]. Periodically, a sample can be taken from each population and frozen, creating a "fossil record" of each evolutionary run.

This kind of "parallel evolution" experiment has many different versions that test for different things. Identical stresses can be introduced into the environment of each population to see how each responds, how evolution solves the problem of survival under stress for each different population.

---

[4]. The Long-term Evolutionary Experiment (LTEE) is led by Richard Linski from Michigan State University.

And what do these kinds of experiments show? Do parallel populations evolve towards very different ends, or very similar ones? Is evolution unpredictable or predictable?

A 2018 paper in *Science Magazine*[5] summarizes the findings from several decades of these kinds of experiments. The authors conclude,

> Although Gould's ideas on contingency have stimulated a great deal of productive work, his view that contingent effects were pervasive throughout evolution remains debatable. The laboratory replays performed to date typically show that repeatable outcomes are common…

This does not settle the debate in an *either/or* fashion. The conclusion is very much *both/and*, that the evolution of life must be *both* contingent *and* convergent. There are things that happen purely by chance and have major influence on outcomes. An asteroid collided by chance with Earth 65 million years ago, eliminating the dinosaurs and opening the door for mammals to become dominant, and that led to humans. But along with the role of chance, evolution is also drawn towards certain outcomes, like the complex eye. It must be concluded that evolution is the interplay of contingency and convergence — we cannot dismiss either.

But how do we explain convergence? What is it that guides independent lineages to the same solutions? What is it that directs evolution, that attracts it to specific outcomes, amidst the chaos of random chance? As yet, we have no answer. The nature and function of DNA itself is far from fully understood; in fact, we've come to realize that we are just scratching the surface. The field of *epigenetics* has revealed that sections of DNA (genes) can be switched off and on as needed in response to environmental changes, opening whole new possibilities for how organisms adapt to the environment, how DNA works, and how life evolves.

---

5. Blount, Zachary D., Lenski, Richard E., Losos, Jonathan B. *Contingency and determinism in evolution: Replaying life's tapes. Science,* 9 November 2018.

Finally, there is an even more fundamental question that modern evolutionary theory has not answered. Why did life evolve beyond bacteria? If survival to reproduce is all that matters, bacteria had that mastered 3 billion years ago. Bacteria are still here today, some in virtually the same form as 3 billion years ago, still surviving better than just about any other life form. Why did life keep evolving into greater *complexity*? The prime directive of Darwinian evolution, survive to reproduce, was fulfilled billions of years ago, and greater complexity was not needed. But life became incredibly more complex over time, eventually resulting in the human brain — the most complex thing we know of in the universe. What, then, makes life move relentlessly towards complexity? Our current scientific theory cannot answer this question. There must be something more to evolution than the Darwin-based modern synthesis tells us. It is, in fact, an incomplete theory, a limited approximation, as all theories in science are.

Creative Emergence

Since Darwin published his groundbreaking theory in 1859, scientists have tended to view life as a kind of machine with discrete parts, and evolution as a machine-like process. This view, known as *reductionism*, was inspired by Isaac Newton's paradigm of the *mechanical universe* that gave rise to classical physics and the Industrial Revolution. But physicists had to radically revise this classical reductionist view of the world starting around 1900, when atomic structure was first probed (more on this in Chapter Five). By about 1930, a completely new kind of physics, called *quantum theory*, was developed that defied common sense, yet gave the right results. So, too, it would seem that biology is in the midst of a radical revision of evolutionary theory and the theory of life itself. Life cannot possibly be a machine because machines can be taken apart into separate pieces and then reassembled to make the same machine. That works for a car, but not for any living organism.

We are beginning to question the belief that the long and improbable journey from bacteria to people was nothing more than a random walk with no purpose. A growing number of scientists have taken a new view of the evolving biosphere and universe characterized by self-organization and creative emergence. Systems theorist Stuart Kauffman writes[6],

> ... The reductionism that has dominated Western science at least since Galileo and Newton ... leaves us in a meaningless world of facts devoid of values. In its place I propose a worldview beyond reductionism, in which we are members of a universe of ceaseless creativity in which life, agency, meaning, value, consciousness, and the full richness of human action have emerged. ... A central implication of this new view is that we are co-creators of a universe, biosphere, and culture of endlessly novel creativity.

Fritjof Capra, a particle physicist turned systems scientist, describes creative emergence beautifully in his recent book[7]:

> Rather than seeing evolution as the result of only random mutations and natural selection, we are beginning to recognize the creative unfolding of life in forms of ever-increasing diversity and complexity as an inherent characteristic of all living systems. Although mutation and natural selection are still acknowledged as important aspects of biological evolution, the central focus is on creativity, on life's constant reaching out into novelty.

The evolution of life is not yet fully understood by scientists, but all science is incomplete. We now expand our look at *evolution* far beyond what happened to life on one planet, as incredible as that was. Let us now explore the evolution of the entire universe and find out how we got an Earth in the first place.

---

6. Kauffman, Stuart. *Reinventing the Sacred*. Basic Books (2008).
7. Capra, Fritjof; Luisi, Pier Luigi. *The Systems View of Life*. Cambridge University Press (2014).

The Evolution of the Universe

You and I are here because of biological evolution, because life got started on this planet and then had enough time to evolve into modern day humans. But life needed a planet to begin with — and not just any planet. It had to be just the right kind of planet, with just the right conditions — a planet with a solid surface, abundant liquid water, enough gravity to hold an atmosphere, and a magnetic field to shield it from radiation, to name only a few. It seems that Earth was perfect. Our planet is a rare paradise, with its many interconnected bio-systems, perfectly balanced yet always changing and surprisingly resilient to the onslaught of our species. How did we get such a perfect planet? Why do planets exist at all? Are there other Earths?

To answer these questions, we must turn to a different domain of evolution that happens at the largest scales of time and space: the evolution of the universe. Astrophysicists have a very good understanding of the evolutionary processes that have been in play since the known universe began and how these processes lead naturally to the emergence of stars and planets. Astrophysicists have a tremendous advantage over other scientists who try to reconstruct the past (like biologists, paleontologists, and historians) because they can literally see into the past and watch it happen.

This nice trick is possible when astronomers use large telescopes to collect light from distant sources (planets, stars, galaxies, and other shining objects). That light, although very fast, takes a while to reach us. Light from the Sun takes about 8 minutes to reach us, so when we see the Sun we are seeing it as it was 8 minutes ago. If it suddenly stopped shining, we wouldn't know it for 8 minutes. Light from the *nearest star,* Alpha Centauri, takes about *4 years* to reach us, so when we look at this star we are seeing it as it was 4 years ago. Most of the stars we see in the night sky are visible as they were thousands or millions of years ago. The farther away a star is, the further in the past we are seeing when we collect its light. With this powerful advantage, astrophysicists have created a very complete and testable explanatory model for the evolution of the universe, or *cosmic evolution*.

The scientific theory of cosmic evolution is called *Big Bang cosmology*. This is the account of how the universe evolved from an infinitely hot and dense point, into the magnificent cosmos we see today, structured with galaxies and stars and planets. The story begins with the unimaginably energetic event we call the *Big Bang*, which was the birth of the physical universe as we know it. What was the Big Bang, and why do we think it happened? And how long ago did it happen, that is, how old is the universe?

How Old is the Universe?

It was 1929 when astronomer Edwin Hubble proposed a completely startling and unbelievable idea. He claimed that the universe is flying apart at enormous speeds, as though a huge explosion took place at some time in the past. He had observed that distant galaxies (1-4), no matter where you look in the sky, are moving *away* from us at very high speeds, and he concluded that the universe is expanding. This dramatic expansion cannot be seen in our everyday lives, or even with small telescopes, but it becomes obvious at the very large distances where galaxies are found — tens and hundreds of millions of light-years away.

1-4

The Pinwheel Galaxy is 21 million light-years away and receding from us at about 150 miles per second. It contains several hundred billion stars, much like our own Milky Way galaxy.

Photo: NASA, Hubble Telescope

Hubble discovered this[8] by gathering the very faint light from galaxies with the Mt. Palomar telescope in California, which was state-of-the-art in the 1920s. He could concentrate the light and then split it into its rainbow spectrum of colors (wavelengths). In that spectrum he found that

---

8. As with many discoveries in science, others contributed to this work, most notably Vesto Slipher and Milton Humason.

the light from every galaxy he studied (about twenty) was stretched out. All of the wavelengths were a bit longer than normal. This stretching of light is called the *redshift*. The redshift happens because the galaxy where the light originates is moving *away* from us at high speed, stretching out the wavelength slightly. Hubble saw redshifted light from galaxies in every direction he looked, and he concluded that all these galaxies were moving *away* from us — never towards us — and that the *farther* away a galaxy was, the *faster* it was moving away from us. This suggests that a massive explosion blew the universe apart long ago, and this has come to be called the *Big Bang*.

From his initial data, Hubble was able to estimate the rate of expansion of the universe, and this rate of expansion is represented by a now-sacred number that astrophysicists call Hubble's Constant, symbolized as H_o. Hubble's Constant is the *spatial expansion rate* of the universe, and once this is known, we can run the expansion backwards in time to the moment when all the contents of the universe were together at one point. This would be the birth of the universe, or the moment of the Big Bang. The very clear expansion of the universe we see at large scales in every direction is the main thing[9] that tells us there must have been a Big Bang, and the rate of expansion that we can measure tells us how long ago the Big Bang happened.

Since about 1930 astronomers have studied thousands of galaxies at increasing distances, out to about 12 billion light-years, with recession speeds approaching the speed of light, confirming Hubble's expansion to a high degree of certainty. It can be shown that the age of the universe is given by the *reciprocal* of Hubble's Constant, and based on today's most accepted value of Hubble's Constant[10], there is widespread agreement that the Big Bang occurred about 13.8 billion years ago. This is the age of the universe.

---

9. There are now many other pieces of evidence supporting the Big Bang.
10. Today's best value of Hubble's Constant is about 72 kilometers per second per megaparsec. Those words after the 72 (called units), may seem weird, but if you unpack their meaning you'll find that they convey the expansion rate of the universe.

It seems that the universe began with a very big bang about 13.8 billion years ago, and it has been evolving ever since.

You may be wondering, what was happening *before* the Big Bang? There are various standard answers that scientists give to this inevitable question, but here's mine: *time itself* was created in the Big Bang, so there can be no *before* if time did not yet exist.

What is the Universe Doing?

If the universe was virtually a point at the moment of the Big Bang, it had to be nearly infinitely hot and dense. From these starting conditions, theoretical astrophysicists have used the known laws of physics to build a mathematical model that predicts what would happen over time as the universe expanded and cooled. These predictions can be compared to actual observations, and if they agree, we might just have a valid model. As of today, most of the theoretical (mathematical) predictions of Big Bang cosmology have aligned very well with actual observations made by astronomers.

When we look at light from the most distant galaxies, we are seeing the universe as it was about 12 *billion* years ago, when the universe was less than 2 billion years old. Wouldn't a biologist love to be able to watch the first animals crawl from the sea to the land? The oldest light we have detected is the *microwave background radiation* that originated when the universe was only about 380,000 years old. Nearly all of our observations of light from the past, and the events it conveys, agree with the predictions of mathematical theory, and this alignment of theory and observation has given us high confidence in Big Bang Cosmology and the story of cosmic evolution. Below is a list of some of the major events of cosmic evolution according to our best science today.

1-5

HIGHLIGHTS OF COSMIC EVOLUTION

EVENT	AGE OF THE UNIVERSE	YEARS AGO
The **Big Bang**	Zero	13.8 billion
Cosmic Inflation, **electrons** emerge	10^{-33} seconds	13.8 billion
The **Four Fundamental Forces** and **quarks** emerge.	10^{-12} seconds	13.8 billion
Protons and neutrons form as quarks join together.	1.0 second	13.8 billion
Nuclear synthesis begins — protons and neutrons fuse to make nuclei of deuterium, helium, and lithium.	3 minutes	13.8 billion
The Great Flash, the universe is cool enough (3,000 degrees) for atomic matter to form, with electrons orbiting nuclei. These highly excited atoms produced intense visible light in a flash that filled the young universe. It lingers everywhere today, stretched into the faint *cosmic microwave background*.	380,000 years	13.8 billion
The **first stars** form. From this time until today, many stars are born, live their lives, and die, creating and distributing all the elements that make up new stars and planets.	200 million years	13.6 billion
Dark matter forms and begins clumping into the large-scale structure called the cosmic web, which becomes the scaffolding for the universe at the largest level.	400 million years	by 13.4 billion
The **first galaxies** form.	700 million years	by 13.2 billion
The rate of **star formation** peaks.	4.6 billion years	9.2 billion
Dark energy overtakes gravity, causing the expansion rate of the universe to begin accelerating.	~8 billion years	~6 billion
Our sun forms.	9.1 billion years	4.7 billion
Planet Earth forms.	9.2 billion years	4.6 billion
The Present	13.8 billion years	Zero

Scientists know far more detail than Table 1-6 contains. In fact, some astrophysicists have devoted entire careers to understanding the first <u>one second</u> in the life of the universe with its preposterous, but apparently necessary, event called *cosmic inflation*. But stepping back from these details, we could say that the main thing the universe seems to be doing is *making stars*. Stars provide everything needed to create the worlds where life can begin and evolve: elements, energy, and planets.

What are Stars?

We have known about stars for as long as people have gazed upwards at the night sky (do other animals do this?), but until recently, no one could explain what stars actually are. In 1838 Wilhelm Bessel made the first accurate distance measurement to a star (beyond the Sun) using the *parallax method*, and the distance he arrived at was unbelievably huge. The nearby star he chose was 60 trillion miles away, or about 10 light-years in today's terms. At such a great distance any light source known at the time would be impossible to see. The source power of this star would have to be a million times more powerful than any process known in the 1800s.

For the next century many more star distances were measured, out to *millions* of light-years, but still no one could explain what it was that powered stars. Finally, in the 1930s, the infant field of nuclear physics provided the answer: it was a *nuclear reaction* that powered stars, specifically a process we call *nuclear fusion*. Nuclear reactions, like fusion, can be a *million times* more powerful than the most powerful chemical reactions we know of, like the combustion of fossil fuels in cars or exploding dynamite.

In principle, stars are quite simple, the perfect balance between gravity always contracting inwards and the outward pressure from nuclear fusion in the core. Stars form in vast clouds, or *nebulae*, of mostly hydrogen (protons) as gravity pulls denser clumps of mass together. Astronomers can see this happening. As the cloud contracts, it begins to rotate, forming a hot central core surrounded by a revolving disk of cooler material. Gravity is the main actor in the formation of stars and the large-scale evolution of the universe.

The central core of the rotating nebula will become a star, and the surrounding disk will turn into a planetary system — this is where planets come from. Planets form *along with* the star, as part of the star. Every star, then, is really a *star system* with an entourage of planets. It would be virtually impossible for a star to form alone without planets, and recent observations of thousands of nearby stars confirm that every star has a planetary system of some sort around it. Because stars are *extremely* abundant in the universe, it means that planets are even more abundant.

As the central core of the "proto-star" keeps contracting, it gets hotter and hotter. When it eventually reaches a temperature of about 15 million degrees, high-speed protons begin to collide head-on and stick together, glued by the *strong nuclear force*. This event is called *nuclear fusion*, and it turns hydrogen (one proton) into helium (2 protons). Fusion reactions release enormous amounts of energy as mass disappears, according to $E = mc^2$.

Once nuclear fusion begins in the core, a stable star is born, and it shines steadily, usually for billions of years. But the life of a star can play out very differently depending on its starting mass. Counter-intuitively, smaller stars live very long lives, and large stars have very short lives. The first 90 percent of the life of any star consists of fusing hydrogen into helium in its core, pouring out prodigious energy, as the Sun is now doing. Stars spend their later life going through increasingly hotter stages of nuclear fusion, manufacturing some or all of the first 26 elements, up to iron. Elements heavier than iron can also be created in low-mass stars like the Sun through more exotic processes. The elements manufactured in low-mass stars are seeded into the universe by stellar winds that constantly flow mass outward from the star. If the star is like the Sun or smaller, fusion must end as nuclear fuel is exhausted. It will gradually cool to become a tiny white dwarf star and finally a cold cinder known as a black dwarf. This will be the Sun's final fate.

When large stars have exhausted their nuclear fuel, they end their lives in a spectacular collapse-and-rebound that triggers a massive *supernova* explosion.[11] In the supernova process, all of the elements can be created and then cast out across the cosmos, to become the raw material for new stars; a tiny, ultra-dense *neutron star* remains, spinning rapidly. The merger of neutron stars, though rare, also produces heavy elements. Figure 1-6 shows the origin processes for the elements in the periodic table, according to the most current theoretical models.[12]

1-6
THE ORIGIN OF ELEMENTS

The colored shading represents the five major processes of element formation. For each element, the base of the box is a time axis, starting with the Big Bang and ending with today.

Courtesy: Chiaki Kobayashi.

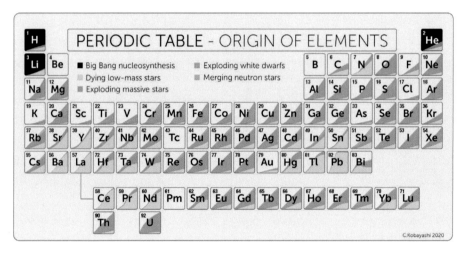

The very largest stars exhaust their nuclear fuel very rapidly and end their lives in even more violent supernovae, leaving a *black hole* behind. Scientists understand the life stages of stars because they can watch all these processes, like star birth, the formation of planetary systems, the stages of fusion, and star death. Each star we see in the sky is in a different stage of its life — some are very young and just forming a planetary system; most

---

11. White dwarf stars can also supernova if they have a companion star to feed on. This is known as a Type Ia supernova.
12. Kobayashi, Chiaki; Karakas, Amanda; Lugaro, Maria. *The Origin of Elements from Carbon to Uranium.* The Astrophysical Journal (September 15, 2020).

are in the hydrogen fusion stage of their lives while others are in their death throes.

We have watched and studied tens of thousands of stars — including our own, up close — and it's safe to say that we understand stars and their evolutionary processes far better than we understand the human body or Earth's climate.

We can now see the three-fold importance of stars in making life on Earth (or anywhere) possible:

- ***Stars manufacture elements.*** The processes within stars and the supernovas that end the lives of large stars create the 92 or so types of atoms we call the elements (Table 1-6). Elements are the building blocks of the world of matter that we know and see and live in, including our own bodies. Astrophysicists like to remind us that the atoms making up our bodies were created by stars, saying "we are the children of stars."

- ***Stars are prodigious energy sources,*** as our Sun is for Earth. Stars are the primary energy source for planets and for living things. Green plants on Earth capture our star's energy through photosynthesis and store it in their hydrocarbon structures. Animals eat plants to derive energy, and the energy contained in fossil fuels, such as coal and petroleum, also originated from the Sun. Most of the Sun's energy is captured by the atmosphere and the oceans, powering the global climate and producing local weather events, such rain, snow, wind, hurricanes, and tornados.

- ***Stars make planets,*** and that's where life might begin and evolve, if the conditions are favorable.

One of the most mind-boggling things about stars is how many there are in the universe. We know that our galaxy has about 10^{11} stars (that's 100,000,000,000 stars), and we know that there are at least 10^{11} galaxies in the universe. This means there are at least 10^{22} stars in the universe. How big is 10^{22}? It has been estimated (more than once) that *the number of grains of sand on all the beaches of the Earth* (try to picture that) is in the neighborhood of 10^{18}. So, there are far more stars in the universe than there are grains of sand on all of Earth's beaches. And every one of those stars has an entourage of planets.

The Big Picture

Let's now step back from details and begin to look at the large-scale arc of cosmic evolution. Very early in its life, the universe began making stars, which was driven by gravitational collapse. Stars form and evolve predictably from well-understood physics, and star-making has continued until today. Making stars seems to be the main work of the universe — or at least the main thing we see going on. We now know with certainty that circling every star are planets. The universe is prodigiously churning out star systems.

The first stars were made almost entirely of hydrogen (with tiny amounts of helium and lithium) because that's about all there was when the universe was about 200 million years old — free protons (also known as hydrogen) and free electrons flying around at very high speeds. But as the first stars were born and lived out their lives, they manufactured elements heavier than hydrogen that were seeded into the cosmos through the supernova process. These elements became the raw material for new stars that could then live their lives, make more heavy elements, and further seed the cosmos. This is the engine of cosmic evolution.

Countless generations of stars have now lived and died, and gradually, more heavy elements have emerged in the cosmos. The Sun, a relative late-comer, is quite special in that it formed with all 92 stable elements[13], and because Earth formed along with the Sun, it also contains these 92 elements. Many

---

13. The Periodic Table now lists an additional twenty five "artificial" elements heavier than Uranium (number 92). These were created in science labs through exotic processes, but they are not stable– they don't exist for long. Most of them decay (turn into something else) in *millionths* of a second. Plutonium, element number 94, is the most stable and is used in nuclear weapons and nuclear reactors.

generations of stars lived and died before the Sun and Earth could form with that full menu of elements.

As the evolving universe has been creating elements through star-making, these elements have found their way into planets. It is here, on planets, where elements can combine into molecules. This is chemistry, as in *hydrogen plus oxygen makes water*. There is very little chemistry happening in or near stars because it's far too hot there — atoms and ions are flying around at very high speeds and can't combine into molecules, but this is not so on planets and other cooler spots. Chemistry, then, becomes an inevitable development in the evolution of the cosmos, and recent observations reveal that complex organic molecules, such as amino acids, can be found in cooler spots throughout the galaxy.

Earth contains the first 92 elements (up to uranium), but oxygen, carbon, hydrogen, and nitrogen are abundant and have combined elegantly and mysteriously to make life. Other less-known elements, like thorium, are critically important in making the Earth biosystem possible. Thorium, for example, powers the radioactive processes that make Earth hot inside (think lava and geysers), and this makes *plate tectonics* possible. Because Earth's crust has floating plates, we have continents among the oceans, plus mountain building, carbon sequestration, temperature regulation, and a host of other essentials that comprise our miraculous biosphere. Earth has the rare combination of a full supply of elements and the ideal conditions for the chemistry of life.

For the physicist, cosmic evolution simply follows the laws of physics, and this quite predictably produces star systems, and some tiny fraction of star systems will have conditions like Sun and Earth. Up until the early 1990s, it was *suspected* that virtually all stars *should* have planetary systems, but no planets had ever been seen because of the enormous distances to stars and the overwhelming brightness of stars compared to planets. But starting in the 1990s, ingenious new techniques have been developed, such as the *radial velocity* method and *transit photometry*, allowing astronomers to detect planets around other stars for the first time. These are called *exoplanets*. We have now observed thousands of exoplanets and have learned two major new things about planets.

First, it is now certain that every star has planets orbiting it — a "bare" star would be virtually impossible because of its strong gravity. All stars, then, are actually *star systems*, like our Solar System. Every one of those twinkling stars in the sky is a dynamic and complex star system with many planets, moons, asteroids, comets, and often a companion star. The second thing scientists have learned recently is that the vast majority of the exoplanets that have been observed appear to be profoundly hostile to life, or at least life as we know it. Many planets are so close to their star that they are roasted by intense radiation. On one recently discovered exoplanet it rains liquid iron! Many other exoplanets are far away from their star and extremely cold, like Pluto. And many are gas giants, like Jupiter and Saturn, with no hard surface at all.

Yet some small fraction of exoplanets are Earth-like — they are rocky and at the right distance from their star to sustain liquid water. A paper published in 2020 estimated that there could be *five billion* potentially habitable rocky planets orbiting Sun-like stars in our galaxy alone.[14] Another paper from 2020 estimated that there should be at least 36 civilizations within our galaxy capable of communicating with us.[15] Yet we have never been contacted, nor have we found any clear evidence of life beyond Earth. It is an open question, whether we are alone in the universe or if life is common. We will return to this question at the end of Chapter Five.

Much like the evolution of life, the universe has been evolving into more and more complex forms. Stars are relatively simple structures, but the heavy elements they manufacture are more complex than the light elements they are born with; atoms of elements combine to make molecules, which can join with other molecules and become the highly complex proteins and DNA that constitute life, and life on Earth exploded in complexity as evolution eventually produced plants and animals and one clever animal

---

14. Kunimoto, Elizabeth, Matthews, Jaymie M. *Searching the Entirety of Kepler Data. II. Occurrence Rate Estimates for FGK Stars*. The Astronomical Journal (May 4, 2020).
15. Westby, Tom, Conselice, Christopher J. *The Astrobiological Copernican Weak and Strong Limits for Intelligent Life*. The Astrophysical Journal (June 15, 2020).

with a very big brain. Through the process of evolution and its relentless drive toward complexity, from the Big Bang until today, the universe has been transformed from hydrogen into humans.

The Evolution of Culture

The evolution of culture is the third great domain of evolution that science has explored. However, the word "culture" is a loaded one, whose meaning is still widely debated among anthropologists, sociologists, psychologists, and philosophers, so I must clarify how I will use the word. Let me begin by saying that culture is a very human phenomenon, and the evolution of culture is the story of the human lineage. More scientifically, we should say it is the story of the genus *Homo*, of which we *Homo sapiens* are the only surviving member. Culture, in the sense I will use the word, is what truly separates humans from all other living things.

There is no consensus definition of culture but Richerson's and Boyd's is a good one:

> Culture is information capable of affecting individuals' behavior that they acquire from other members of their species through teaching, imitation, and other forms of social learning.[16]

We could simplify this even further to say that *culture is knowledge that is transmitted from one individual to another*. Humans are very good at this — we talk and write to each other, we explain and teach and learn. We role model, tell stories, make movies, write poetry, give lectures, perform music, create paintings, and more. All of these transmit knowledge and meaning. This is culture. But is this really unique to humans? Don't other animals also transmit knowledge and learn from each other? Just a few decades ago most scientists thought that only humans were capable of culture, but now we have many documented examples of animal culture (I will use "animal" to mean non-human animal).

---

16. Boyd, Robert, Richerson, Peter. *Not By genes Alone: How Culture Transformed Human Evolution.* University of Chicago Press (2005).

Many mammals and birds teach their young how and where to find food. Chimps and gorillas have complex hierarchies and customs within their social groups. Orcas (killer whales) hunt in groups by creating rings of bubbles that enclose schools of fish, a technique that young whales learn from their elders. Ravens and chickadees each use at least 100 different expressions that convey information. Whales and elephants are well-known for their long-distance communications, though we don't know what they are saying to each other.

Indeed, many mammals and birds are capable of this kind of simple culture. But culture was taken to a whole new level in the human lineage, far beyond what any other animals are capable of. What is it that makes humans, and our hominin[17] ancestors, different from all other animals? What does it mean to be human? In the story of the evolution of life on Earth, when did humans diverge from the rest of life to eventually become the near-geologic force we now are?

Based on the most current findings in paleoanthropology, we can now place the birth of our lineage to be roughly 3.5 million years ago in Africa. This marks the split in the evolutionary road that led to modern humans, and in the bigger story of the evolution of life, it was a very recent development. In the 24 hours representing the age of the Earth (Table 1-2, page 2), this was about *one minute* before midnight. But what was it, exactly, that happened in Africa 3.5 million years ago to launch the journey of humanity?

The answer is that certain ape-like hominins that we call Australopiths, who had begun walking upright and who had hands with opposable thumbs and dexterous fingers, began manufacturing and using simple tools made of stone. No other animal had ever done this before. The oldest manufactured stone tools are about 3.3 million years old and were found in Ken-

---

17. I will use the term *hominin* to refer to our ancestors, such as *Australopithecus* and *Homo erectus*, who were walking upright and using their dexterous hands for new purposes but were not yet truly "human." I prefer to reserve the term *human* for our own species, *Homo sapiens*, although anthropologists vary in what they call human.

ya.[18] This find represents the emergence of an entirely new capability, and an entirely new regime of life, that became the genus *Homo*, our lineage.

Lest you protest that other animals use tools, it is true that many animals do use simple tools from natural sources — chimps use sticks to extract termites, sea otters use rocks to pry loose and break open abalone shells, octopuses gather coconut shells to make shelters, birds build nests, and capuchin monkeys use two stones as a hammer and anvil to break open nuts, a skill that reportedly takes eight years to master. But none of these animals, or any animal, can consciously and intentionally *manufacture* tools from carefully chosen and sourced materials, using precise sequences of steps and high-level skills. Perhaps even more importantly, our ancestors were able to teach each other this sophisticated craft and pass along this knowledge to others, so that it survived and grew over many lifetimes.

This was a new kind of culture in which extensive knowledge was transmitted over many generations, and innovations were continually added, so that the tools kept improving. This can be seen clearly in the archaeological record. The evolution of culture is preserved over millions of years in the increasingly sophisticated stone tools that paleoanthropologists have collected from ancient sites in Africa and around the world. With the first manufactured stone tools, our ancestors had moved into a new realm of cognitive function that has continually evolved until today.

Let us now distinguish between "animal culture" and the "hominin culture" that is unique to our lineage. Cognitive psychologist Merlin Donald distinguishes between these by using the term *episodic culture* to refer to the rudimentary culture of animals that culminates with the well-documented behaviors of apes and the term *mimetic culture*[19] to refer to the "true culture" that emerged in Africa when our ancestors first began manufacturing stone tools. From now on I will use the word "culture" to mean this higher-level culture. In Chapter 4 we will further explore the cognitive roots of the cultural revolution that uniquely defined our lineage.

---

18. Harmand, Sonia, Lewis, Jason E, et al. *3.3 million-year-old stone tools from Lomekwi 3, West Turkana, Kenya*. Nature (May 20, 2015).
19. . We further explore the work of Merlin Donald in Chapter 4.

Paleoanthropologists have been excavating and cataloging stone tools and skeletal remains (teeth survive better than anything else) from all over the world for more than a century. Stone tools are durable and can survive intact through millions of years, unlike the flesh of the body, and they tell a clear story of human evolution. One of the major challenges facing scientists is dating, that is, finding out how old an artifact is. Many ingenious dating methods[20] have been discovered and continue to improve, giving us a well-evidenced chronological picture of the progression of tools representing the evolution of culture.

It is now widely thought that the very first tool-makers and users were members of the genus *Australopithecus*. But the period of time between about 3.5 and 2.5 million years ago is sparsely documented, and paleoanthropologists are not in full agreement on names for the various species found from this time. However, everyone agrees that by about 2.5 million year ago a new hominin with a growing brain and improving tools had emerged. This was no longer an Australopith and has been given the genus name *Homo*.

The first member of the *Homo* lineage is widely known as *Homo habilis*, although alternate names have been proposed. But everyone agrees that by about 2.0 million years ago a very distinct species called *Homo erectus* was roaming Africa and soon spread into Eurasia and Asia. *Homo erectus* was more human-like and had developed a new generation of tools, and many new behaviors, such as hunting and fire-making. Figure 1-7 depicts the mainstream view today of the many species of both *Australopithecus* and *Homo* that lived and died before us today. The exact line connecting *Homo habilis* to *Homo sapiens* is still unclear, but we know more all the time about some of our fascinating cousins, like the Neanderthals. Every one of these species shown in 1-7 is now extinct, except for modern *Homo sapiens*, appearing as the tiny black box at the top right of the chart. In the next section we will explore in more depth the story of humanity.

---

20. Such as radioactive decay (numerous methods), thermoluminescence, electron spin resonance, paleomagnetism, and dendrochronology.

1-7
CHART OF HUMAN EVOLUTION

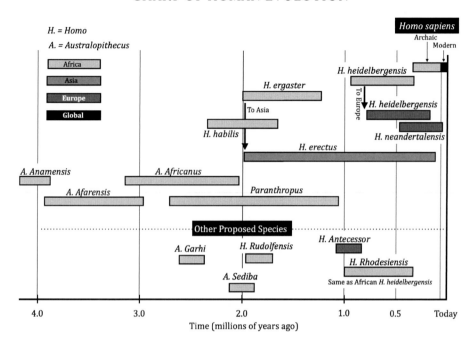

The Story of Genus Homo

The Australopiths lived 4.3 to 1.1 million years ago. During this time, at least seven species of *Australopithecus* roamed Africa. They were our most immediate ancestors and had mastered upright-walking and developed increasingly dexterous hands. Two different species, *Australopithecus africanus* and *Australopithecus afarensis,* are proposed to be the first toolmakers and the ancestral link to the *Homo* lineage. The brain volume of the *Australopiths* eventually maxed out at about 500 cubic centimeters (cc). By comparison, modern chimps have a brain volume around 350 cc, while modern humans reach about 1400 cc. This lineage continued for more than a million years after the divergence of *Homo*, and the later forms are now called *Paranthropus*. The entire lineage eventually went extinct.

First Stone Tools (Around 3.3 million years ago). The oldest manufactured stone tools come from East Africa and date to between 3.3 and 2.8 million years old — for simplicity, we'll call this 3 million years ago. The first generation of stone tools are called *Olduwan* tools since they were first found at Olduvai Gorge in Kenya. Figure 1-8 shows some typical Olduwan tools, and, although they may look crude, they required a high level of skill and planning to manufacture.

1-8

TYPICAL OLDUWAN TOOLS FROM ABOUT
2.5 MILLION YEARS AGO

Stone tools were constructed by using two stones — a hammer stone and a core stone — one in each hand. But these had to be made of the right *kind* of stone. The core stone, typically flint or chirt, had to be softer than the hammer stone. Toolmakers not only had to learn what stone-types were correct, but they also had to find sources for these materials and pass along this knowledge to others. With precisely directed blows, sharp-edge flakes could be stripped from the core stone to shape it precisely for its intended use. Both the sharp-edged flakes and the core stone were effective for many purposes, including better preparation of plant and animal food sources and to chop and shape wood.

In tool construction, the angle of the blows had to be correct and the proper sequence of steps had to be learned, so that the second round of strikes would sharpen rather than dull the first-struck edge. These skills and procedures required extensive practice and rehearsal before they yielded finished Olduwan-style tools. Olduwan tools are found in many parts of

Africa, spanning a period of more than a *million* years (until about 2 million years ago), and they gradually became more sophisticated as new innovations were contributed to the tool culture.

All of these unprecedented skills and procedures that toolmaking required were communicated in a new way within family-based communities. However, this transmission of knowledge extended over large geographical distances. The art and science of toolmaking spread throughout Africa as mimetic culture caught fire. A new cognitive mode called *mimesis* had emerged for the first time on Earth (Chapter Four).

By about 2.5 million years ago a new hominin species emerged in Africa that was distinctly different from any *Australopith*. For one thing, its brain volume (cranial volume) reached 600 cc, surpassing any previous hominins. Several species names have been proposed for the first member of *Homo*, however the name *Homo habilis*, "handy man," is most widely used.[21] Evidence in the fossil record of *Homo habilis* is scant, but it probably weighed up to 100 pounds and stood up to five feet tall. Its brain volume of about 600 cc is generally taken as a primary defining criterion for the genus *Homo*. For the next 2 million years, the brain volume in the *Homo* lineage continually increased, eventually reaching 1600 cc for Neanderthals and 1400 cc for modern humans. None of the *Australopiths*, or any of the other great apes, experienced this phenomenal brain growth. It is unique to the *Homo* lineage, but no one knows why this happened. Meat eating, enhanced by tool use, has been suggested as a contributing factor. *Homo habilis* eventually went extinct about 1.7 million years ago, but meanwhile, a new species of *Homo* emerged in Africa. This was *Homo erectus*, the most successful of all our ancestors.

Homo erectus (2.0 million years ago to 50,000 years ago) was lean and tall, more intelligent with a brain volume eventually reaching about 1100 cc, and was a smooth runner. A new generation of tools appears with *erectus* — these are known as *Acheulean* tools (Figure 1.9), after a site in France where they were first found in the 1800s. *Homo erectus* lived in nuclear

---

21. A parallel species, *Homo rudolfensis*, has also been proposed, but evidence is scarce.

families and mastered fire use and hunting. It is unclear whether *erectus* descended directly from *Homo habilis* or from another species of *Homo,* but not long after *erectus* appeared in Africa, this new species traveled to Central Asia — about 1.8 million years ago — and eventually into China and Southeast Asia. *Homo erectus* was the first of our ancestors to leave Africa.

1-9
TYPICAL ACHEULEAN TOOLS FROM ABOUT 1.5 MILLION YEARS AGO

The original members of this species who stayed in Africa are often referred to as *Homo ergaster,* while the more widespread Asian members are called *Homo erectus* (and some scientists simply use the name *erectus* for both the African and Asian populations). This species had a long and successful run throughout Asia before finally going extinct, perhaps as recently as 50,000 years ago, in Southeast Asia.

<u>Homo heidelbergensis</u> (approximately 1.0 million years ago to 200,000 years ago). Back in Africa, *Homo ergaster* was morphing into a new species that began to look more like modern humans — anthropologists call this species *Homo heidelbergensis. Homo heidelbergensis* had a brain volume reaching 1250 cc (compared to our maximum of about 1400 cc) and developed a wide range of more sophisticated tools and behaviors, including spear points, permanent shelters, burial of the dead, elaborate nature rituals, and vocal communication (though not yet truly speech and language).

Some members of *heidelbergensis* were able to leave Africa, and some stayed; this was an important parting of ways. The ones who left showed up in Europe about 800,000 years ago, and they spread everywhere from England and Germany to Spain and Greece. From these *heidelbergensis* populations in Europe and Eurasia, several new species diverged around 400,000 years ago, most notably the Neanderthals (*Homo neandertalensis*). By about 200,000 years ago, the remaining members of *heidelbergensis* disappeared while the Neanderthals, and some cousins like the Denisovans, were top dogs throughout Europe and Eurasia.

Homo sapiens (Starting around 300,000 years ago). While the Neanderthals ruled the roost in Eurasia and Europe, with a brain growing up to 1700 cc, back in Africa *heidelbergensis* was gradually transitioning into a new species that looked anatomically very much like us today. If you cleaned up one of the males and put him in a suit, he would blend right in on Wall Street, but he wouldn't be able to speak in words and sentences. Anthropologists call this new species *Homo sapiens idaltu*, meaning archaic *Homo sapiens*; they are commonly referred to as *anatomically* modern humans, but not yet *behaviorally* modern humans.

For the next 200,000 years, these archaic humans struggled to survive massive droughts in Africa. Starting between 150,000 and 100,000 years ago, two separate groups of people successfully moved out of Africa. One group moved east from the Horn of Africa into the Saudi Peninsula, and the other group traveled north through Egypt and into the Levant (modern day Israel). It was in the Levant, around 100,000 years ago, that humans first encountered the Neanderthals, who were at the southern reach of their Eurasian territory. Skeletal remains found in Israel dating to around this time show both human and Neanderthal characteristics, suggesting that the two species interbred. For more than a century it was assumed that Neanderthals and humans were different species (hence the different species names) and would not be able to interbreed. But now, there is much rethinking about what defines a species and the relationship between humans and Neanderthals.

No one knows what the encounter was like when humans and Neanderthals met for the first time, but we must suspect that it didn't go that well for the humans. The tools of the two species at this time, known as

Mousterian tools, were remarkably similar, suggesting an equivalent level of intelligence, but humans were smaller in stature and had a smaller brain. The humans may have been overpowered and driven back, or simply killed, but recent genetic evidence suggests that some mixing of genes also took place (yes, sex). Whatever it was that happened, signs of human presence in this part of the world completely disappear for the next 40,000 years, so it seems that this first attempt by humans to leave Africa by traveling north ultimately failed. Apparently, they couldn't get past the Neanderthals in the Levant. However, this first encounter with the Neanderthals was only round one for humanity.

Other groups of people exited Africa further south between 150,000 and 80,000 years ago, crossing the Straits of Hormuz at the Horn of Africa to enter the Saudi Peninsula. They probably followed coastlines as far as India and China. Human teeth have been found in China dating to about 100,000 years ago,[22] but the prevailing evidence now suggests that none of these Asian populations survived for the long-term. There are several lines of evidence suggesting that by about 70,000 years ago the only surviving human population on Earth was located in Africa.

Near-extinction and The Great Leap (73,000 to 60,000 years ago). Both genetic and skeletal evidence suggest that all of us today — all living humans — can trace our ancestry to Africa around 60,000 years ago. This would have to mean that the populations of people who left Africa between about 150,000 and 75,000 years ago, probably in multiple waves, did not ultimately survive and contribute to the modern human genome. Recent DNA studies of humans living everywhere on Earth suggest that all modern people originated from two small populations in Africa about 60,000 years ago.[23] One group lived in East Africa (modern day Kenya and Ethiopia), and nearly all modern humans are descendants of this group. The other group lived in South Africa, and they became the modern Khoisan Bushmen.

---

22. Callaway, Ewan. *Teeth from China reveal early human trek out of Africa*. Nature (14 October, 2014).
23. Behar, Doron, et al. *The Dawn of Human Matrilineal Diversity*. American Journal of Human Genetics (May 2, 2008).

It appears that the entire population of humans on Earth living between about 70,000 and 60,000 years ago plummeted to perhaps a few thousand, and they all lived in Africa. What caused this near-extinction of humans? One explanation, first proposed by Stanley Ambrose,[24] puts the blame on the catastrophic volcanic eruption of Mount Toba on the island of Sumatra about 73,000 years ago. The magnitude of this eruption was thousands of times greater than Mt. St. Helens in 1980. It was a global crisis, triggering a volcanic winter that may have lasted for thousands of years. This event is well documented in locations around the world, with 30-foot layers of volcanic ash in India and drastic changes in plant life in the Middle East.

In the immediate aftermath of the Toba eruption, the people who had left Africa and moved into the Saudi Peninsula, India, and China were in the direct line of fire, downstream in the prevailing winds. Apparently, none of these people survived. Some may have survived the initial catastrophe but not the drastic change in climate that followed. However, the people who lived in Africa were geographically safer and less exposed to the immediate devastation of the eruption. They managed to survive. After 3,000 generations, all of us today are related to each other through this small group of Africans.

While the human population on Earth was decimated by the Toba event, the other two dominant *Homo* species at the time, *Homo erectus* and *Homo neandertalensis*, were largely spared the brunt of the Toba eruption because they lived in Asia and Europe, more favorably situated geographically. But, in the end, the real winners were the *Homo sapiens* in Africa. This crisis of near-extinction and extreme environmental stress apparently catalyzed an evolutionary transformation in humans that catapulted them into a completely new stage of cognitive development. Neither *erectus*, nor the Neanderthals, nor of any other species, experienced this dramatic cognitive leap.

---

24. Ambrose, Stanley. *Late Pleistocene human population bottlenecks, volcanic winter, and differentiation of modern humans.* Journal of Human Genetics (June 1998).

The archaeological record suggests that the humans in Africa who survived the Toba event acquired a whole new suite of abilities that their ancestors did not have — nor did the Neanderthals in Europe. These more advanced humans would eventually become a geologic-scale force that overwhelmed all other living things, altering the face of the Earth. This dramatic transformation that humans experienced is referred to as the "Great Leap" by Jared Diamond in *Guns, Germs, and Steel* (1997). Stanford University anthropologist Richard Klein calls it the "Big Bang of consciousness" in his 2002 book, *The Dawn of Human Culture*. Paleoanthropologists refer to this as the transition from the *Middle Stone Age* to the *Late Stone Age,* signifying the emergence of *behaviorally* modern humans with the scientific name, *Homo sapiens sapiens.*

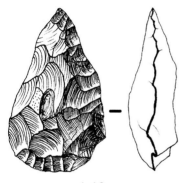

1-10
Typical Hand Axe,
Front and Side View.
Public Domain/WikiMedia

The Great Leap was a turning point when humans became an unstoppable force on Earth, and they soon spread to every habitable continent, driving many other species to extinction. The evidence showing this transformative leap is dramatic in the archaeological record. Before the Great Leap, tool manufacturing had slowly but steadily progressed, with tools becoming more sophisticated over several million years. For most of this time the primary tool was the hand axe (1-10), and later, spear points and crude blades appeared. Although the craftmanship was very impressive, there were not that many different *kinds* of tools. About 100,000 years ago, humans in Africa and Neanderthals in Europe and Eurasia had very comparable tools, referred to as the *Mousterian* tool industry, and they shared advanced behaviors, such as burial of the dead, primitive art, symbolic beads, and long-distance trade.

But starting around 60,000 years ago there was an explosion in the human tool set, and not in the Neanderthals'. In a relatively short period of time, many new innovations appeared in human settlements, such as fishhooks, sewing needles, harpoons, engraving tools, tempered micro-blades, drilling and boring tools, musical instruments, carved figurines, and elaborate cave art (1-11). Something had happened inside the human brain, inside the human, but no one knows exactly what that was. The volume of the human brain had reached a maximum of about 1400 cc by at least 300,000 years ago, so the Great Leap was not due to an increase in brain *size*. Apparently, some kind of internal reorganization took place.

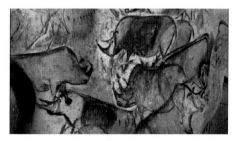

1-11
A 32,000 year-old cave painting of rhinos from *Chauvet Cave in France.*
Public Domain/Wikimedia

The New Humans (emerging around 60,000 years ago). Before the Great Leap, humans had struggled to leave Africa, and those who did apparently did not survive, perhaps victims of the Toba event and its aftermath. But the new humans who emerged from the Great Leap moved easily out of Africa and spread over the entire planet. By 50,000 years ago, they found their way into Australia; by 40,000 years ago, they entered Europe, and by 20,000 years ago, they had moved across Siberia and were crossing the Bering Straits (a land bridge at that time) into North America.[25] By about 15,000 years ago, most of the modern United States was occupied by people we now call Native Americans, and some even made it all the way to Southern Chile by this time. These "super-humans" who came through the Toba bottleneck had capabilities far beyond any of their *Homo* ancestors, or any other living thing in the history of Earth.

When fully modern humans arrived in Central Asia and Europe around 40,000 years ago, they once again encountered their old rival, the Neanderthals. The details of this second meeting remain mysterious, just like the

---

25. A team of researchers recently reported that human tools found in Northern Mexico are more than 30,000 years old, but this remains unconfirmed and controversial. Ardelean, Ciprian, et al. *Evidence of human occupation in Mexico around the Last Glacial Maximum.* Nature (July 22, 2020).

first meeting 100,000 years ago, but we do know a few things for sure. First, all evidence of the Neanderthals disappears shortly after this. And second, the two species interbred (yes). This was discovered in 2003 when DNA was extracted from a Neanderthal finger bone (a remarkable accomplishment) and analyzed, revealing the first glimpse of the Neanderthal genome. After comparison with human DNA, it was found that modern humans of European descent possess a small amount of Neanderthal DNA. Since then, additional samples and better methods of genome sequencing have shown that all humans today possess traces of the Neanderthal genome,[26] making it certain that humans interbred with Neanderthals somewhere along the way, and probably in two different periods of encounter.

Beyond this we don't know why the Neanderthals disappeared — whether humans simply outcompeted them for resources, killed them off, absorbed them into the human genome, or some combination of these. But one way or another, these new humans who made it through the Toba bottleneck thrived while the Neanderthals languished and disappeared, probably because they had not experienced the developmental leap that humans did. Round two went to the humans, and it was game over for the Neanderthals.

Many anthropologists suspect that one of the unique abilities humans acquired at the time of the Great Leap was the mastery of spoken language. The acquisition of speech and language required both the proper vocal anatomy and major new cognitive abilities for processing large numbers of symbols (words) and creating meaning from them. These abilities may have been emerging for hundreds of thousands of years before the Great Leap, but around 60,000 years ago humans must have reached a level of sophistication in spoken language that allowed them to collaborate in powerful new ways. They could work and problem-solve together, greatly amplifying the intelligence and capabilities of a single individual. These fully

---

26. People of European descent have the largest amounts of Neanderthal DNA, presumably because their direct ancestors mated with Neanderthals. But modern Africans and Southeast Asians also have traces of Neanderthal DNA (but less than Europeans) because some of the early humans with Neanderthal DNA migrated *back* to Africa and into Asia carrying pieces of the Neanderthal genome that now linger. Ackey, Joshua, et al. *Identifying and Interpreting Apparent Neanderthal Ancestry in African Individuals.* Cell (January 30, 2020).

modern humans who emerged from the Great Leap eventually developed an extensive oral tradition of storytelling and mythology, the so-called *mythic culture*.

The Built World: (Beginning about 11,500 years ago). For the entire story of the *Homo* lineage so far, our ancestors were nomadic hunter gatherers who lived intertwined with nature, leaving very few signs of their presence. But starting around 11,500 years ago, as the last Ice Age was ending and Earth entered a warmer period, something unprecedented appears prominently in the archaeological record: *megalithic structures*. These are large stone monoliths arranged and stacked intentionally. The oldest known megalithic structures are at Göbekli Tepe in modern-day Turkey, dating to about 11,500 years ago, and are claimed to be the first temples in the world.

Over the next few thousand years, megalithic structures were constructed at many locations around the world. Perhaps the most famous and iconic megalithic structure is Stonehenge in England, which is thought to be about 5,000 years old. It is a mystery how people quarried, moved, and stacked stones weighing as much as 80 tons; and perhaps, even more puzzling is how this happened at widely separate locations around the world among people who had no direct contact with each other.

Before about 11,500 years ago, there is no sign of anything like this anywhere in the world. Göbekli Tepe and other megalithic sites signal the beginning of the *built world*, or we could say the "man-made" world in older parlance. The appearance of the built world marked the end of a major era and the beginning of a new one — the one we still live in today.

As Earth entered a warmer period, the Middle East was lush and fertile, and people living there began planting and tending simple gardens as a food source. This practice, called horticulture, required consistent tending and protecting, which was not compatible with a nomadic lifestyle. People began to put down roots in permanent settlements, so they could tend their gardens; domesticated animals, like cats, dogs, and sheep, were a natural.

By about 10,000 years ago, large permanent settlements of people began to appear in the Middle East at sites such as Jericho (the modern Palestinian Territories) and Çatalhüyük (modern Turkey). These are referred to as *proto-cities* and were the precursors of the true cities that emerged a few thousand years later. The advent of proto-cities was a profound change in the human condition. It was the end of the ancient nomadic lifestyle in which people were embedded in nature, and the beginning of a new sedentary lifestyle that fundamentally disconnected people from nature. People now lived *inside* dwellings with walls and ceilings that separated them from the natural world. Yet this was only the first step towards a completely new social arrangement that was still a few thousand years away: *civilization*.

In the first proto-cities, like Jericho, people lived in mud-walled structures packed together side-by-side with an entrance in the ceiling and a ladder down into the dwelling. Archaeologists have found that all these dwellings were very similar, and there were no large buildings. This suggests very little social stratification, no rich or poor people, no royalty or slaves. These proto-cities show almost no signs of *social hierarchy*. This conclusion is further supported by analyses of skeletal remains that show very similar levels of nutrition among all the residents — apparently no one ate much better than anyone else. These societies were largely egalitarian, and probably very chaotic, with no authority figures running things. The abundance of female figurines found at these sites also suggests that these societies were largely matriarchal.

By about 7,000 years ago the first true cities appear in the archaeological record. They were originally built around temples. Among these were Uruk and Eridu in Mesopotamia (modern Iraq). Starting around 5,000 years ago, monumental public architecture such as palaces, temples, and pyramids, appears for the first time. This required well-organized construction projects, powered by armies of slaves. Someone was in charge while many others did the work. This signifies the beginning of social stratification and the top-down power structures needed to maintain the social hierarchy.

Now, for the first time, there emerged emperors, kings, and high priests: the all-powerful ruling elites. This was the beginning of *civilization*.

These first cities had public spaces, palaces and temples, well-planned streets, water and sewer systems, and separate districts where different social classes lived. By about 5,000 years ago, some cities grew to acquire territory around them, to become *city-states*; now for the first time, we find fortified walls around cities and armaments for war. Territory had to be taken, to be conquered, so now people had to defend themselves from attack or, perhaps, attack first and take possession. *Empires* soon emerged, and the current era of history began. US versus THEM would be the recurring theme for the next 5,000 years, right up until today.

Many important innovations accompanied the advent of civilization around 5,000 years ago, including the invention of writing, large-scale agriculture, metallurgy (the end of the stone age), standing armies and military operations, institutionalized religion (often a part of state power), property ownership (by those in power), laws, government, taxes, currencies, trade networks, specialized professions, male dominance of the public sphere, and the list goes on. While none of these things existed just a few thousand years earlier, this is the civilizational paradigm we still live in today.

The Great Transformations

Let's now step back from the fascinating story of the human lineage, the story of *Homo*, and recognize three major cultural transformations, each one launching our ancestors into entirely new regimes of existence. These three transformations stand out from anything else in the archaeological record over the last 3 million years. To appreciate this, we just need to look at what things were like for our ancestors before and after each transformation. How did they live, and what were they doing, before and after? How big of a change did the transformation represent? In each of these transformations, the change was dramatic and launched a new stage of evolutionary development.

1-12

THE GREAT TRANSFORMATIONS IN THE *HOMO* LINEAGE
(Not to Scale)

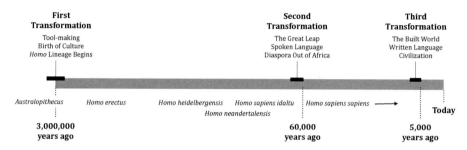

The First Transformation (around 3 million years ago). The significance of this transformation is not only that a few individuals began making and using stone tools, but that this practice spread so widely across Africa and the tools improved continually in both craftsmanship and utility. A knowledge base extending beyond one lifetime emerged for the first time on Earth. This is the essence of culture because culture *is* transmitted knowledge. For the first few million years of hominin culture, knowledge could not yet be transmitted through the use of words or language as we know it. Knowledge was transmitted through mime, gesture, embodied imitation, and skill refinement. Merlin Donald calls this suite of new abilities *mimesis*. The mimetic capacity gave rise to two-way communication in a mime-like fashion. As far as we can tell, no other living thing has acquired the mimetic ability with two-way expressive outputs.

The Second Transformation (between 73,000 and 60,000 years ago). This transformation is a textbook example of an evolutionary leap stimulated by extreme environmental stress — a near-extinction event, a *population bottleneck*. The people who survived the Toba event had enhanced abilities and new powers, probably the result of new cognitive capacities that supported the mastery of sophisticated spoken language. It is now suspected that the Neanderthals had the vocal equipment that was necessary for speech but never acquired the cognitive abilities to support complex language. The humans who emerged from the Great Leap bore almost no resemblance in their capabilities to those who lived before. These new humans spread quickly over all the habitable continents of Earth, and we have dominated the biosphere ever since.

The Third Transformation (between 11,500 and 5,000 years ago). Before this transformation, all the people of Earth were nomads who lived closely connected with nature. By the time this transformation was completed, some 5,000 years ago, many people were living in busy cities, inside buildings with rectangular spaces and under the rule of a king or emperor. Social hierarchy, patriarchy, monumental architecture, war and empires, government, laws, monetary systems, writing, professions, metal technology, agriculture, and the wheel — all of these appeared on Earth for the first time as part of the third transformation.

Culture and Consciousness

In this chapter we've toured the three great domains of evolution that science has explored: the evolution of *life*, the evolution of *the cosmos*, and the evolution of *culture*. However, we should not think of these as three distinctly separate phenomena, but as different expressions of the same evolutionary impulse that is found everywhere, driving a relentless creative movement towards complexity, diversity, and higher functioning — *Supra-evolution*. We are beginning to see the universe as not simply a collection of objects in space, but as an *evolutionary process*.

The evolution of human culture is recorded clearly in the stone tools that our ancestors left behind, and even today our tools (now "technology") speak volumes about us. Over the span of 3 million years, we see evidence of the steady evolution of humans into more complex social networks and more sophisticated behaviors, punctuated by occasional developmental leaps that we have called the great transformations. What was it about our ancestors that was evolving? It was not simply their physical bodies because we know that in the last 300,000 years the human form has not changed appreciably, including the size of the brain. So, it must be that something *inside* was evolving.

Apparently, some sort of rewiring or reconfiguring has been going on *within* that big brain, bringing new cognitive abilities, such as mimesis and language. Yet scientists know very little about how the physical brain actually produces thoughts, ideas, feelings, and images or how we create and accumulate knowledge; we have no information about the actual brain structures of our ancestors because the delicate tissues of a brain cannot be preserved over time, like stones and bones can.

To understand the evolution of culture, and to make sense of the human story, we must move beyond merely trying to understand the physical brain, to something much bigger and deeper associated with the brain, something completely mysterious, yet totally obvious — *Consciousness*. This one word is the most loaded symbol known to humans, representing something we all experience at every waking moment, yet something we cannot measure, or touch, or see, or explain. Despite this difficulty, *consciousness* will become the central construct of this book from here forward, and for that reason, the next chapter is devoted entirely to exploring the centuries of effort and the many ideas proposed to understand consciousness. We will not end up with a neat and tidy definition because consciousness is not a *thing* that can be located or measured with an instrument, but we *will* develop a working knowledge and an operational sense of the word that we can use throughout the rest of the book.

I must now introduce the critically important relationship between *culture* and *consciousness*. Although they are not the same thing, we will see repeatedly that the two are inextricably intertwined. When culture first emerged on Earth millions of years ago, it began to co-evolve with consciousness in a synergistic, positive feedback loop, launching our lineage onto the spectacular trajectory that led to modern humans. Let us for now, describe the close relationship between culture and consciousness in this way:

> Culture is the *exterior* expression of consciousness.
> Consciousness is the *interior* basis for culture.

Consciousness is what underlies and gives rise to culture. Culture is what shows up in the world. As goes one, so goes the other. When we see evidence of a transformative leap in human culture, it means there has been a transformative leap in human consciousness. The evolution of culture is simply a reflection of the *evolution of consciousness.* Consciousness, and the fact that it evolves, is the very thing that has made humans different from all other living things on Earth. We cannot begin to understand the phenomenon of humanity, or our story, without understanding the evolution of consciousness. It is behind everything that humanity has done, or is doing, or can do in the future.

CHAPTER 2

Consciousness 101

… consciousness implies a continuous (but interruptible) stream of phenomenal senses or experiences — a technicolor, multimodal, fully immersive and wholly personalized movie, playing to an audience of one.

— Anil Seth

As the human lineage evolved over the last 3 million years, the brain *tripled* in size, for reasons unknown, while our closest living relatives, the chimps and bonobos, had no such thing happen to them. We are truly the animals with the big brains. But while this physical brain was expanding in size, something else was emerging in humans, something that doesn't show up directly in the fossil record, on any instruments, meters, or measuring sticks: a growing *self-aware consciousness* — the fully integrated sensual, emotional, cerebral experience of being YOU, of being an embodied "I" in the world, a person with a life history and the possibility of conscious choice, creativity, and agency. Consciousness is pervasive and ubiquitous, like water for a fish, but the word remains difficult to define and nearly impossible to explain. It is the greatest mystery there is — the mystery of *being* — and yet we experience it, we are "in" it, at every waking moment. Our highly evolved consciousness is what sets humans apart from all other life forms on Earth; it's what gives us the tremendous power we have in the world.

The remainder of this book deals with consciousness in one way or another, so we must take time in this chapter to develop a sense of the domain it represents. This chapter provides an overview and a sampling of an immense body of knowledge from many fields of study, and it is not possible in a few thousand words to do justice to all that has been written and said on the subject of consciousness. Scientific research into consciousness is in its infancy, and scientists are far from understanding the role of the physical brain in producing consciousness; not to be deterred by the magnitude of this undertaking, let's jump right in!

SOME PROPOSED DEFINITIONS OF CONSCIOUSNESS

- *The having of perceptions, thoughts, and feelings; awareness.*

  – Macmillan Dictionary of Psychology

- *Awareness by the mind of itself and the world.*

  – Oxford Living Dictionary

- *A state of mind in which there is knowledge of one's own existence and the existence of surroundings.*

  – Antonio Damasio, from *Self Comes to Mind*

- *Wakeful presence.*

  – Jean Gebser, from *The Ever-Present Origin*

Consciousness in Philosophy

Philosophers have a long history of trying to understand the nature of the mind, or the psyche, going back to Classical Greece. Only in the last century or so have people begun to make the connection between the mind and the physical organ we call the brain. Today, most of us take the connection between the mind and the brain to be a given ("use your head!"; "she's so brainy"), and most scientific research focuses on trying to explain the relationship between the non-material mind and the material brain. But science is still far from such an explanation.

A fundamental and thorny question for philosophers concerns the relationship between matter and mind, between the world "out there" and the world within. One of the first modern philosophers to tackle the "mind-matter" question was Rene Descartes (1596 — 1650), and his answer has come to be known as *Cartesian dualism*. He proposed two distinctly separate domains of existence: an outer *objective* world of material things and an inner *subjective* world of mind and personal experience. In Descartes' dualism, the brain (specifically the pineal gland) was the interface, a kind of portal between these two worlds, between mind and matter, the subjective and the objective, the inner and the outer.

For most of us Cartesian dualism *is* our waking experience as a body with a mind. When we are awake, we experience our body and ourselves participating in an outside world of objects ("objective reality") that seems to be independent of us. This is the "real world," a shared reality that we all agree upon remarkably well. All of us who are functioning normally see the same bus coming down the street and know not to step out in front of it.

But inside each of us is our private world of subjective experience that we alone can know. If you and I both look at a red rose, I can never know what *your experience* of the red rose is like or how it compares to *my experience*, yet we will both agree that we are seeing a red rose. Philosophers call discrete conscious experiences, like the experience of seeing the red rose, *qualia*. Qualia are "what it is like" or "the way it feels" to have mental states, such as feeling pain, seeing the color red, or smelling a rose, but they are entirely subjective, private experiences that only *you* can have. I can't feel your pain or experience your smell of the rose. No one knows what qualia actually are or what is happening in the brain to produce this element of subjective experience.

But there's an 800-pound gorilla in this philosophical room: subjective experience is owned and felt by *someone*, and that someone is *you*. There is a personal *self* behind consciousness, the protagonist (or victim) in every experience, the thinker of thoughts, the doer of deeds, the feeler of feelings. You smelled the rose, and you are the one and only star of that *phenomenal* experience. The nature of *selfhood* or *personhood* is a critically important matter that has perplexed sages, saints, philosophers, and scientists alike. There is no consensus about what the self is, and some have even questioned whether it exists at all. The emergence of the self, the *self-aware* part of consciousness, is one of the most important developments in the ongoing story of evolution. We will return to the question of the *self* many more times in this book.

After Descartes, until today, philosophers of mind have continually debated between *dualism* and *monism*. As we have seen, dualists like Descartes say there are fundamentally just two things that exist, two distinctly different substances that make up the world: mind and matter. But monists say there is just *one* thing, and then they differ as to what that one thing is. There is the *mind* camp of monism championed by *idealists,* like Immanuel Kant (1724 — 1804). They say that there is *only mind*, and the physical world is created by, or is an aspect of, mind. The *matter* camp says there is *only matter* (which now includes energy) and matter creates the mind, that is, the brain creates the mind. Most scientists today take this view. Then there's the *neither* camp of monism saying that neither mind nor matter are fundamental, but both are the manifestation of something even more fundamental, and that is the *one thing* of monism. Yet no one agrees about what that fundamental *one thing* actually is.

Over the last few centuries the matter camp of monism has prevailed, and this view now underpins our entire modern worldview, our reality: the world is made of matter — objects in space, whether they be atoms or buses or people — and mind is a product of the atoms and molecules of the brain. This philosophical view has come to be called *materialism* or *physicalism,* and it is the default position in science today. It is fair to say that a large majority of practicing scientists believe that consciousness can be explained by physical interactions of matter — in the brain, of course. Virtually all scientific research on consciousness today, largely in the field of neuroscience, is seeking to find out how the material brain creates the mind and other facets of consciousness.

One of today's best-known materialist philosophers is Daniel Dennett. He argues that consciousness as a "thing" is an illusion, and all that's going on is molecules of the brain obeying the laws of physics and chemistry. Dennett and supporters rightly point out that the history of science is full of phenomena that could not be explained and were thought to be magic of some sort until materialist explanations began to prevail in the eighteenth century. In the late 1700s light was widely considered supernatural and beyond scientific explanation, and electricity was believed to be an invisible massless fluid. Then, we discovered electromagnetic waves and the electron. For Dennett there's nothing mysterious about consciousness — it's what physical brains do.

At the other end of the spectrum are philosophers like David Chalmers and Christof Koch, who have suggested that consciousness may be a fundamental characteristic of the universe, like mass, space, or electrical charge. The view that everything has consciousness to some degree — a rock, a cell phone, our planet, or the whole universe — is called *pan-psychism*. Proponents of this view include Plato, Spinoza, Leibnitz, and William James, and this view is also found in Eastern traditions like Taoism and Buddhism, so Chalmers and Koch are not in bad company.

Koch and Giulio Tononi have proposed the *integrated information theory*, which says that any system could be considered conscious if the information it contains is sufficiently interconnected and organized. The best example of this kind of system is the brain, but the internet also qualifies, and so do personal computers. Which leads to the inevitable question: will artificial intelligence (AI), now housed in computers, ever be able to attain the equivalent of human consciousness, or even surpass it? Most AI engineers say yes, but many others have their doubts.

Many of today's philosophers agree that there are two broad types of consciousness: *p-consciousness* (or phenomenal consciousness), which gives rise to our immediate raw experience of awareness and qualia, and *a-consciousness* (or access consciousness) that calls up stored information from past experiences to create the autobiographical experience. In principle, a-consciousness is easy to explain in mechanistic terms because it resembles the way computers access and use memory. But explaining p-consciousness, and how the immediate experience of awareness arises from the brain, is often referred to as the "hard problem" — a term Chalmers coined in 1994.

Other philosophers contend that there are more than two types of consciousness. William Lycan identifies at least eight types including organism-, control-, introspective-, subjective-, and self-consciousness. Other philosophers propose even more, so you can see there is no real consensus. Consciousness does not easily succumb to these categories with labels that philosophers have tried to establish. Despite centuries of philosophical discussion, we have made very little progress in understanding consciousness.

Consciousness in Psychology

By the late 1800s, psychology diverged from philosophy as a distinct field of science through the pioneering work of Wilhelm Wundt in Germany, Ivan Pavlov in Russia, and William James and Sigmund Freud in the U.S. But over the last century the study of consciousness and the field of psychology have been deemed "soft" science because numerical data are so hard to acquire. The physical sciences have it easy because there are so many ways to get data from the physical world, so many instruments that spit out numbers. But in psychology it's impossible to measure mental states or consciousness with any ruler, meter, or instrument.

In lieu of instrumentation, psychologists have relied heavily on studying human subjects through their observed behaviors and verbal accounts — anecdotal evidence — in the attempt to understand the structure and operation of the mind. If done carefully, interviews of subjects can be a powerful source of information, even though it is not "hard data." Psychology attempts to identify, map, and understand the mental structures that determine various behaviors and functions, such as cognition, thought, attention, image-making, memory, perception, language, emotions, and metacognition (thinking about thinking).

In the early days of psychology, the most fertile sources of insight into the mind and brain were cases where things went very wrong. Perhaps the most famous of these cases is that of Phineas Gage, who had a steel shaft driven completely through his skull in an accidental explosion at a railroad construction site (2-1)

Amazingly, Gage stood up and walked with assistance shortly after the accident and lived for another twelve years. He had sustained severe damage to many parts of his brain, yet he was eventually able to function in a surprisingly normal fashion — except that his personality was transformed from a sober, responsible, hardworking man into that of a cheating, cursing, lying gambler with a volatile temper.

2-1

Sketch from 1868 by John Harlow, Phineas Gage's doctor.

Public Domain

This was a rather crude way of learning about the brain, but it revealed several new insights. First, it showed that the brain is not just a collection of parts (like a machine), but rather a holistic unit with parallel structures that can compensate for damage. Secondly, it showed that personality and character traits are correlated with the brain, something that was not understood before this. Many other cases of unfortunate accidents causing significant injuries to the brain have brought new insights and strengthened the correlations between brain anatomy and behavior.

Another direct source of knowledge about the brain comes from open brain surgery, where the patient is awake and locally anesthetized with an exposed cortex. When doctors use a tiny electrode at a small voltage to gently stimulate particular areas of the exposed brain, the patient reports experiences of smell, sound, visual landscapes, and feelings — all in a dreamlike experience. This leaves no doubt that the physical brain has a major role in creating the experiences we have and our experience of the world. But how does it do this? We don't know.

Another case where things have gone wrong is mental illness. Psychologists, and more recently psychiatrists, have sought to identify and treat brain illnesses, such as severe depression, bipolar disorder, and schizophrenia, but we have made very little progress in understanding and effectively treating these debilitating disorders. We do know that pharmaceuticals and natural psychoactive drugs can have profound effects on consciousness and the conscious experience. Unfortunately, the lab-synthesized pharmaceuticals used by many psychiatrists have been only modestly successful in treating mental illnesses or furthering our understanding of the mind. These molecules that alter brain chemistry often cause side-effects and unintended consequences that have to be further treated with other drugs.

However, the so-called *psychedelics* seem to hold much more promise. These include psilocybin (derived from a mushroom), mescaline (from the peyote cactus), DMT (from a shrub called ayahuasca), and LSD (originally derived from a mold). In the 1960s, Timothy Leary and Richard Alpert, two psychologists at Harvard, experimented enthusiastically with LSD and theorized that it held the key to unlocking the mystery of schizophrenia and consciousness itself. They took a lot of it and encouraged *everyone* to take a lot of it. They were soon fired from Harvard, but not before LSD became a staple of the counterculture. As a result, the U.S. government soon banned research into psychedelics, and this lasted for the next forty years. Research is now resuming, with keen interest in psilocybin and low doses of LSD. The plant-derived psychedelics could be powerful tools in the future for understanding consciousness and treating mental illness. For much more, see Michael Pollan's *How to Change Your Mind* (2018).

Merlin Donald, a cognitive psychologist from Queens University, Ontario, proposes that *conscious capacity* has increased steadily throughout the evolution of higher life (the last 500 million years), reaching sophisticated levels in birds and mammals and going off the charts in humans. Donald proposes a number of *skill clusters* that comprise conscious capacity.[27] The high-level skill clusters listed below make up what he calls the *executive suite* of conscious capacity that humans have. These are some of the things that contribute to human consciousness and set humans apart from other living things.

---

27. Donald, Merlin. *A Mind So Rare*. W.W. Norton & Co (2002).

1. *Self-monitoring of success or failure.* All Great Apes (including humans) are capable of this, but not rats or cats.

2. *Divided attention*, or multi-tasking. Spoken language requires this because speakers must keep track of their own behavior and that of others. Most apes are not capable of this.

3. *Self-reminding.* This is necessary in maintaining long sequences of actions, for example the construction of tools. Bonobos and enculturated apes are capable of this, but few other animals.

4. *Self-recognition.* The mirror test is used to uncover this: a subject is placed in front of a mirror, then the mirror is removed and a red dot is discretely painted on the subject's forehead. Then the subject is again placed before the mirror. All humans and some apes will recognize themselves and touch the dot on their forehead, but monkeys and most other animals fail to recognize themselves. However, a gorilla in captivity named Coco was taught simple sign language and was then observed signing to her roommate, "Coco big beautiful mountain gorilla," a clear sign of self-awareness.

5. *Whole-body imitation.* This is the mimetic ability to mirror another person, first in body movements, then in dance, language, and other behaviors. The emergence of this cognitive skill about 3 million years ago enabled the manufacture and use of stone tools and ignited the transmission of culture. Some enculturated apes are capable of this, but not even at the level of young children.

6. *Mindreading.* This is the capacity to understand that there are other minds and that they are predictable. Every "A" student uses this skill effectively to figure out what their teacher wants, what they are thinking. Most apes are minimally able to do this.

7. *Pedagogy.* This is the ability to teach and learn. This is the foundation of culture, and it was highly amplified by the emergence of language. Only humans are capable of this.

8. *Symbolic invention.* Apes and young children can use rudimentary hand gestures and expressions to represent ideas, but apes are very

limited in their ability to express themselves. Spoken and written language, allowing conversation, teaching, and learning, is far beyond the capability of any other animal.

9. *Assembling complex skill hierarchies.* Driving a car requires learning how to start the car, how to turn, back up, accelerate, shift gears, read signs, monitor other cars, and so on. All non-humans are very limited in this capacity.

Consciousness in Neuroscience

Neuroscience is the most recent field of science that seeks to understand consciousness, and its holy grail is to find the connection between the physical brain and the conscious experience, or consciousness. This quest is based on *epiphenomenalism,* the assumption that consciousness is a by-product of the functioning of underlying physical structures in the brain, and it is confined entirely within the brain's processes.

Neuroscientists want to know how the brain produces the full experience of being alive and awake as a human being with a mind and a self. To do this they must bring together knowledge from many areas of science, including neural anatomy and physiology, psychology, medicine, and philosophy. In recent decades, they have made more progress than anyone else in explaining the brain and consciousness. One of today's leading neuroscientists and writers on the subject is Antonio Damasio, Professor of Neuroscience, Psychology, and Philosophy at the University of Southern California and the Salk Institute. His 2010 book, *Self Comes to Mind,* is a tour de force in explaining neuroscience research, and I will draw heavily from it in the following discussion.

The neuroscientist straddles the two worlds of Descartes: the physical world of brain anatomy and physiology and the non-physical world of mind and conscious experience. Rapid advances in knowledge have been made on the physical side with new imaging techniques, and we can now correlate many cognitive and emotional experiences, or mental states, with specific regions of the brain. Imaging techniques, such as the functional

MRI (fMRI), give us a real-time movie showing blood flow in the brain where neural activity is higher. When a subject is experiencing a particular mental state, such as fear, or is engaged in some cognitive process, such as problem-solving, a specific part of the brain lights up on the fMRI. From these kinds of studies, we have created a complex 3D map of brain regions and smaller centers, generally called nuclei, which are correlated with different aspects of consciousness, such as emotion, vision, hearing, reasoning, imagining, thinking, and awareness. But beyond these correlations, we still have very little idea how the physiological processes happening inside the brain give rise to the many facets of consciousness.

Looking at the brain from the physical side, its gross anatomy is simple — 3 pounds of warm, wet, spongy tissue taking up about 1400 cubic centimeters of space, divided into two outer hemispheres — right and left — and three main structures nested from the inside out — the brain stem, the thalamus, and the cerebral cortex. The brain stem is the core of the brain, sitting on top of the spinal cord, and is thought to be the oldest and original brain structure that first emerged in reptiles, so it is popularly called the reptilian brain. It manages the many automatic tasks of *homeostasis* in the organism — maintaining the narrow range of perfect conditions everywhere inside the organism. It is also responsible for many other things, including the feeling of embodiment and wakefulness, the "fight or flight" instincts, and other typical reptilian capacities.

The thalamus serves as a way station for information collected from the body, via the brain stem, and destined for the cerebral cortex. All signals bound for the cortex stop at thalamic relay nuclei to be rerouted to their specific destinations in the cortex. The thalamus also talks back and forth with the cortex in recursive loops that integrate information. It relays critical information to the cortex and massively inter-associates cortical information. Only smell escapes a stop through the thalamus by traveling directly from the nose to the cortex.

In contrast to the brainstem, the cerebral cortex evolved most recently along with the thalamus. It is the outermost region of the brain and is divided into a right and left hemisphere. The cortex is larger (thicker) proportionately in humans than in other apes, mammals, or any living things. As the brain tripled in size during the evolution of the *Homo* lineage, it was probably the cerebral cortex that was growing. In interplay with the brain stem and thalamus, the cortex constructs the maps and images that become mind. Undoubtedly, this is where thinking takes place, although we don't understand what that actually is. Its vast memory banks record our experiences in the physical and social environments we live in and from this construct our autobiographical self. The cortex is the final stop in the globalized interplay of information that produces our conscious mind.

According to Damasio,[28] *assembling the consciousness show is such a cooperative effort it would be unrealistic to single out any particular partner. Autobiography could not arise without the seminal contributions of the brain stem toward the protoself, or without the brain stem's obligate consorting with the body proper, or without the brain-wide recursive integration brought in by the thalamus.*

This means that we cannot take a mechanistic view of the brain as being made of discrete parts that each perform a certain function. It works as a *whole*, with massive interplay among many structures.

The brain is made almost entirely out of two kinds of cells, called *neurons* (there are many types) and *glial cells*. Neurons and their axons are embedded and suspended in a scaffolding made of glial cells that provide both physical support and nourishment. But it is the neurons that do the business of the brain. With about a hundred billion of these all wired together electrically, the brain is the most complex thing we know of in the universe.

Within each major region of the brain there are smaller centers, or nuclei, that are heavily interconnected and these nuclei have interconnected substructures, which have interconnected substructures until we reach the level of the single-cell neuron. This massive interconnectivity happens through

---

28. Damasio, Antonio, *Self Comes to Mind*. Vintage Books (2010).

the electrical "firing" of neurons. When neurons fire, a small voltage pulse is generated and sent down the length of an axon that connects through a synapse to the dendrite of another neuron, or network of neurons, to make a microscopic electrical circuit.

According to Damasio, the firing of a micro-neural circuit, triggered by some input such as the sensing of red light, is probably the origin and seed of a conscious event like a particular mental state or subjective experience. These seed events or *protophenomena* are then scaled up across a nested hierarchy of structures throughout the brain, integrated with many other proto-events, and sent to the cortex to create the full-scale images, feelings, and thoughts that comprise conscious experience.

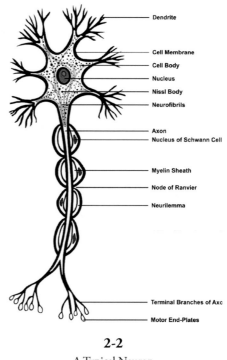

2-2
A Typical Neuron

The prevailing view within neuroscience is that consciousness emerges from the firing of neurons in the brain, which is a completely classical, or mechanistic, view. When neurons fire, they are on, and when they are not firing, they are off. This on-or-off state of neurons seems to be a binary language of zeros and ones, just like computers that use on-or-off switches (transistors) to represent zeros and ones. Because there are about 100 billion neurons in the brain that can be interconnected in a nearly infinite number of ways, it seems plausible that the brain could be a binary computer with enough capacity to produce the conscious awareness we experience. But no one is close to explaining what is happening in my brain when I have the experience of smelling a rose or when I imagine a giant redwood tree in my mind.

The brain-as-computer model *can* account for what has been called the "easy problem" of non-conscious functions of the brain — the "auto-pilot" or "zombie" modes that regulate the many body functions and responses. A thermostat is the simplest kind of "automaton," consisting of a temperature sensor and a switch that goes on and off at a chosen value of the temperature. Brain circuits can do this using the on-off firings of neurons, and today's computers do this with transistors. Researchers have now built sophisticated robots and artificial intelligence systems (AI) that mimic some brain functions very well. It has now been conceded that a human cannot come close to competing with a computer in the game of Chess or in the much more complex game of Go. Artificial intelligence enthusiasts feel that it is only a matter of time before AI can achieve full consciousness and surpass humans altogether.

But not so fast. The "hard problem," as David Chalmers called it, is how the brain produces phenomenal conscious experiences, subjective feelings, and self-awareness. How does the brain produce the conscious experience of "qualia," such as our experience of the color blue or the smell of the ocean? The computational model of the brain, with neurons firing, has no answers and further fails to address these other questions:

- How does the brain "bind" together the many neuronal activities into a unified conscious experience that persists over time?
- What is perception? We understand how senses collect information and send it to the brain, but how does the brain create the picture of the world that we actually experience?
- How do neuronal firings produce the *gamma synchrony*, a 30 to 90 Hz oscillation, that shows up in EEGs and correlates with consciousness?
- How do single-celled organisms swim, find food, learn, remember, and do many other sophisticated things, *without neurons*?

The mind by itself would be merely a set of dazzling routines, like a movie showing at an empty theater, without being owned and experienced by a *self*.

Who is it that experiences these products of the mind, and where does he/she/it live? People have long believed that the self was a little person, or *homunculus*, residing somewhere inside the head, but dissection of the brain reveals no such thing. In Damasio's view the self is built in three stages, as follows:

- The *protoself* emerges from information in the brain stem about the internal state of the organism and generates the feelings of a living body, so called primordial feelings. The protoself is the timeless experience of *being a body*.

- The *core self* is generated by the interaction of the organism/protoself with an object. The images of organism and object are modified and linked in a coherent pattern to produce a narrative sequence — a drama of the organism in the world, experienced in short vignettes. In the core self, an episodic sense of time now exists.

- The *autobiographical self* arises from the integration of images from the core self with stored memories of past experiences, or anticipated future events. This creates the full experience of being *you*, a human being with a body, a world around you, a personal life story, and perhaps future aspirations — and conscious self-interest. An outer region of the cortex, known as the *posteromedial cortices*, is strongly implicated in the construction of the autobiographical self, supporting the view that only humans have reached this stage of selfhood (sorry, not your dog).

We can see that the self should not be viewed as a thing but as a collection of *processes* — more accurately, a system of integrated processes spread throughout the triad of the brain stem, thalamus, and cortex. But we are far from understanding the details of how the neuro-electrical maelstrom of brain activity creates the self. According to Damasio's *psychology of the self*, it is simplistically made of four integrated image elements or aspects:

- **Primordial feelings** signify the existence of *your* living body.
- **Perspective**, or viewpoint allows you to see and understand things from *your* point of view.

- **Ownership** is the feeling that your mind belongs to *you* and no one else.

- **Agency** is being able to carry out actions chosen and commanded by *you*.

The self, whatever it is, owns a mind. In neuroscience and cognitive psychology, the *mind* is the massively interconnected set of processes culminating in the cortex and creating images and maps, perception, memory, thinking, feeling, reasoning, language, knowledge, understanding, and learning. The mind, in this sense, is the toolbox of mental abilities that the modern human brain brings to its owner.

That owner of the mind is the interconnected system of brain processes that comprise the self: the experiencer, the knower, the protagonist, the witness, the chooser, the thinker, and the one who owns and uses the capabilities of the mind. In this view, the mind is a set of tools that the self can use, akin to a computer that needs a user. The user is the self. It is when the *self* and the *mind* combine — when *self comes to mind*, as Damasio says — that consciousness arises. In other words,

$$mind + self = consciousness$$

This should not be taken like 2 + 3 = 5, but rather that consciousness is the highly entangled marriage of mind and self.

Consciousness in Physics

Before about 1925, the concept of consciousness was not within the boundaries of physics. The physics of this early era, referred to as *classical physics,* was established by Isaac Newton (1643-1727) and further developed by many other geniuses through the 1700s and 1800s. Classical physics was (and still is) spectacularly successful at describing the world that we live in and know through our senses. It spins-off powerful technologies, like the steam engine, the telegraph, the airplane, and spacecraft, proving its validity and power. By about 1880, many scientists felt that classical physics was infallible.

In classical physics, the scientist — also known as "the observer" — is *outside* and *independent of* the thing being observed, whether she is watching the moons of Jupiter through a telescope or witnessing a compass needle respond to a nearby electrical current. The observer, and her consciousness, plays no role in the experiment. Obviously, the act of looking at one of Jupiter's moons does not have any effect on that moon, so this makes perfect sense. Therefore, consciousness is more than irrelevant in classical physics; it is not even a consideration. It does not exist. As Descartes spelled out, there is a world of matter and a world of mind, and the two are completely different.

However, when physicists began exploring the sub-microscopic world of the atom in late 1800s, they quickly found that classical physics could not explain the behavior of atoms, particularly the way they absorbed and emitted light. This became a true crisis in science for the next few decades and required an entirely new kind of physics. Through the combined efforts of some of the greatest physicists and mathematicians of the twentieth century, including Max Planck, Albert Einstein, Niels Bohr, Werner Heisenberg, Erwin Schrödinger, Wolfgang Pauli, and Paul Dirac, this project of reinventing physics was completed by about 1927 and is now known as *quantum theory* or *quantum physics*. And while classical physics ignored the role of consciousness, quantum physics confronted it.

In the sub-microscopic world of quantum physics, the mere act of observing something changes it. We see an object because photons of light bounce off of it and enter our eye. Photons of light have no effect if we are observing a large object like a table, but if we are observing an electron, a collision with a photon is patently disruptive. At the quantum level, the act of observation changes the thing being observed. To put it another way, the presence of consciousness and how it interacts with the world (by collecting photons, for example) changes the world. This now puts the observer *in* the experiment, not outside looking in as in the classical view. Quantum theory attempts to include the observer, and consciousness in general, in its descriptions of the world. It makes consciousness a fundamental aspect of the "physical world," blurring the clean separation between matter and mind (sorry Descartes). *Idealists*, such as Kant, took this to the extreme, saying that consciousness is *all* there is and matter is an illusion (or a manifestation of consciousness).

Henry Stapp, a renowned quantum physicist and philosopher at Lawrence Berkeley Labs, states that "the most radical change wrought by this switch to quantum mechanics is the injection directly into the dynamics of *certain choices made by human beings about how they will act.*[29]"

He is referring to the observer's choice of what to observe and how to observe it. The observer narrows the scope of the inquiry to some small slice of the world, a process called *partitioning*, and this act, this free choice by a human, plays a significant role in the outcome of any experiment or measurement at the quantum level. Stapp further elaborates that:

> What the founders of the new theory claimed, basically, is that the proper subject matter of science is not what may or may not be "out there" … it is rather what we humans can know and can do in order to know more. Thus they formulated their new theory … around the knowledge acquiring actions of human beings. The focus of the theory was shifted from one that basically ignored our knowledge to one that is about our knowledge, and about the effects of the actions that we take to acquire more knowledge upon what we are able to know.[30]

One of those founders of quantum theory, Werner Heisenberg, put it this way:

> The laws of nature which we formulate mathematically in quantum theory deal no longer with the particles themselves but with our knowledge of the elementary particles. The conception of objective reality … evaporated into the … mathematics that represents no

---

[29]. Stapp, Henry. *Mindful Universe: Quantum Mechanics and the Participating Observer*, Springer (2007).
[30]. Ibid.

longer the behavior of elementary particles but rather our knowledge of this behavior.[31]

Quantum theory describes a system, such as an atom or molecule, using a mathematical equation known as the *Schrödinger wave equation,* or the *wave function* for short. According to Nobel laureate Eugene Wigner, "given any object all possible knowledge concerning that object can be given as its wave function"[32]. The wave function requires a user, an agent, who makes choices about what to measure or observe. We can ask it what a hydrogen atom will be like under certain specific conditions, and it will give us probabilities. It does not say where an electron will be located at some instant as it orbits the nucleus of an atom, but rather, it gives a probability of the electron being found in some region of space. The orbits of electrons in atoms are "probability clouds," something that chemistry students are told.

The Schrödinger equation gives only *potentialities* — the many possible states of a system — until it *collapses,* or *reduces,* to an *actuality*. According to the "Copenhagen interpretation" provided by Bohr and Heisenberg, this collapse of the wave function occurs when a measurement is made on the system by an observer, who ultimately is a human with consciousness. Heisenberg wrote in 1958:

> The observation itself changes the probability function discontinuously; it selects of all possible events the one that has actually taken place. Since through the observation our knowledge of the system has changed discontinuously, its mathematical representation has also undergone the discontinuous change and we may speak of a "quantum jump".[33]

This has been misinterpreted popularly to mean that things do not become "real" until they are observed, sometimes called the "observer effect." But

---

31. Heisenberg, Werner. In *Daedalus*, Volume 87, Number 4. The MIT Press (1958). Quoted by Eugene Wigner in *Remarks on the Mind-Body Question,* an essay in *Symmetries and Reflections*. Indiana University Press (1967).
32. Wigner, Eugene. *Symmetries and Reflections: Scientific Essays*. Indiana University Press (1967).
33. Heisenberg, Werner. *Physics and Philosophy: The Revolution in Modern Science*. Harper (1958).

"real" things presumably existed before human observers were around. Is a tree real if no one is looking at it?

And what about robots, without consciousness? Can a robot be an observer? Some human, *with* consciousness, designed and built the robot to make a chosen observation or measurement, and the human will retrieve and examine the data from the robot, so human consciousness is clearly required before knowledge is possible. This means that a robot is still just a *tool* used by human consciousness, like a microscope or thermometer. What is important in quantum theory is that there will be an *increase in knowledge* that results from the observation and the collapse of the wave function, and knowledge implies human consciousness.

Quantum theory does not explain and describe the objects of the world as classical physics does, but rather, it is a theory of *what is possible to know* about the world. And most importantly, for our purposes, is that the increase in knowledge from a reduction event (an observation) is acquired by *someone* with *consciousness*. In classical physics consciousness does not exist, but quantum theory brings it front and center. The human observer makes a free choice about what to measure and how to measure it, and as a result he gains an increment of knowledge. In the classical world the observer is an invisible voyeur, playing no role in the outcome of an experiment or observation, but in the quantum world the observer is *part of* the experiment and cannot be ignored.

The quantum world is fuzzy and interconnected, unlike the classical world of hard objects separated by space. The full nature of the relationship between consciousness and the world of matter is still widely debated among quantum theorists, but it is clear that the neat separation of mind and matter, posited by the dualists like Descartes, does not hold at the most fundamental levels of reality.

Quantum theory acknowledges consciousness, but can physics explain what consciousness actually is and how the brain and body conspire to produce it? Since the 1990s, Nobel laureate Roger Penrose and his collaborator, Stuart Hameroff, have been looking for answers at a much deeper level of brain anatomy — in the vast system of *microtubules* that underlies all cellular structure, including neurons. Microtubules are hollow cylindrical polymers of the protein tubulin that perform many important functions in cells. They are major components of the cell cytoskeleton that gives cells their shape; they manage chromosomes during cell division; they transport materials, and they bundle together to make cilia, flagella, and other structures that produce locomotion.

It was proposed in 1973 by biologist Jelle Atema that microtubules in the brain might also process information.[34] Stuart Hameroff, a medical researcher at the University of Arizona, discovered in the 1980s that microtubules could function as "molecular automata" with bit-like switching, like a simple computer microprocessor.[35] With linkage to neighboring tubulins, he realized that a massive amount of computing power was possible. Physicist Roger Penrose learned of Hameroff's work and proposed that microtubules were the site of *quantum* computations, not merely classical, binary computations.

Penrose and Hameroff began collaborating in the 1990s and have become the champions of the view that the microtubule system is the foundation of consciousness. Each neuron in the brain has about 100,000,000 (10^8) microtubules within it. And since there are about 10^{11} neurons in the brain, it means there are about 10^{19} microtubules in the brain. Now, we can say that's a *very big* number, but we get a better feel for just how big that is by recalling that this is roughly the number of grains of sand in all the beaches on Earth. That's about how many microtubules there are in the human brain.

Hameroff found that microtubules can vibrate at a rate of about 10^7 cycles per second (10 MHz), and with 10^{19} microtubules capable of doing that, we can estimate the total computational capacity of the brain to be about 10^{26} operations per second. Researchers in AI have made similar estimates for the much bigger and clunkier neuronal structure of the brain,

---

34. Artema, Jelle. *Microtubule theory of sensory transduction.* Journal of Theoretical Biology (1973).
35. Hameroff, Stuart. *Ultimate Computing.* Elsevier (1987).

and they arrive at about 10^{16} operations per second, a miniscule fraction of what the microtubule system can achieve. Today's best computers have now reached this level (10^{16} Hz), and some AI theorists have suggested that this should be enough to approximate human consciousness. In other words, we are not that far off from creating consciousness artificially. But the much higher capacity of the microtubule system deep within the brain pushes that AI prediction much farther away. In fact, no computer today comes close to approximating human consciousness.

Up until recently neuroscientists and AI engineers assumed that it was the interconnected neurons of the brain and their rapid firing that gave rise to the mind and consciousness. But now it appears that the microtubule system underlies the neuronal system and feeds information up to it. Operating at the heart of the microtubule system is quantum physics because the sub-microscopic size of the microtubule (25 nanometers across) is at the quantum scale while neurons are not. According to Penrose and Hameroff, quantum computations occur within microtubules, culminating with an *objective reduction* (OR) event, represented by the collapse of the wave function. These OR events in the microtubules are coupled with higher-level structures, such as dendritic gap junctions, that trigger neuronal activity and, perhaps, give rise to a "conscious moment." EEG data from the brain show a distinct gamma wave synchrony at a frequency that averages about 40 Hz and correlates with conscious activity. This suggests that consciousness might stream at about 40 conscious moments per second.

Another piece of the picture is the quantum entanglement of many microtubule structures. Entanglement is a prediction of quantum theory that has also been observed experimentally in recent decades. Very small systems of particles, such as electrons, protons, and small atoms, can become interconnected (or *entangled*) across empty space, even across large distances. Penrose and Hameroff have proposed that a conscious moment originates when billions of microtubules become entangled and act together, not as classically separated objects but as something more like Jell-o that wiggles everywhere when poked in one place.

In an essay from 2017, Penrose and Hameroff sum up their *Orch OR* theory of consciousness as follows[36]:

> We proposed in the mid 1990s that consciousness depends on biologically 'orchestrated' quantum computations in collections of microtubules within brain neurons, that these quantum computations correlate with and regulate neuronal activity, and that the continuous Schrödinger evolution of each quantum computation terminates in ... 'objective reduction' of the quantum state (OR). This orchestrated OR activity (Orch OR) is taken to result in a moment of conscious awareness and/or choice. This particular form of OR is taken to be a quantum-gravity process related to the fundamentals of space-time geometry, so Orch OR suggests a connection between brain biomolecular processes and fine-scale structure of the universe. Orch OR places the phenomenon of consciousness at a very central place in the physical nature of our universe ...suggesting that conscious experience itself plays a role in the operation of the laws of the universe.

Here they are suggesting that consciousness originates even more deeply than the microtubule system of the brain, in the very space-time fabric of the universe. This is entirely speculative, but their suggestion aligns with the Buddhist view that consciousness is not limited to the brain. Physicists like Heisenberg, Wigner, Penrose, Stapp, and other quantum theorists have explored the question of consciousness deeply, yet physics, and all of Western science, remains far from a complete explanation.

Consciousness in Buddhism

While Western scholars have been struggling for the last 400 years to understand the nature of mind and consciousness, scholars in India started on this quest at least 3000 years ago. The *Upanishads*, a series of texts written in Sanskrit starting about 700 BCE and part of a larger body of work

---

[36]. From Chapter One of *Consciousness and the Universe: Quantum Physics, Evolution, Brain, and Mind.* Edited by Penrose, Hameroff, and Subhash Kak. Cosmology Science Publishers (2017).

known as the *Vedas*, contain the world's first map of consciousness, and they provide the core concepts of both Hinduism and Buddhism. Buddhism originated in India around 450 BCE with the life of Siddhārtha Gautama, also known as *the Buddha*, and it spread throughout Southeast Asia, taking on many different forms or schools. Buddhism is not a religion, per se, but it is a philosophy and a psychology of the world that aims to alleviate suffering and elevate consciousness. It has no concept of God in the way most religions do, and no missionaries to spread the word.

By about 700 CE, Buddhism found its way into Tibet, where it came to be known as *Vajrayana*, and it was here that the study of consciousness proliferated. Today's head of Tibetan Buddhism is the wise and charming Tenzin Gyatsu, also known as the Fourteenth Dalai Lama, who fled Tibet in 1959 after his country was seized by the Chinese government. Since then, he has made his home in Dharmsala, India and has been a frequent visitor to the U.S., where he participates in discussions about consciousness, such as the recurring *Mind and Life Conference*, with some of the leading neuroscientists, philosophers, physicists, and psychologists in the world.

There are several striking differences between the Western and Eastern approaches to consciousness. First, the word "consciousness" has no direct translation or equivalence in the Sanskrit language native to Buddhism — there are at least five different words that stand for the components, or aspects, of consciousness. A further difference is that Western scholars are narrowly focused on the brain and its role in consciousness, while in Buddhism there is virtually no interest in the brain as the creator of consciousness.

The primary tool for studying consciousness in Tibetan Buddhism is contemplative practice, or meditation. For many centuries, Tibetan monks have devoted their lives to contemplative practices, along with intense peer discourse, and through this process they have extensively explored and mapped the inner landscapes of consciousness. In the Western view, consciousness disappears in dreamless sleep, in a coma, or under anesthesia — it's either on or off — but in the Eastern view consciousness is never gone. It just has deeper levels.

According to Indian philosophy, consciousness is that which is *luminous* and has the capacity for *knowing*. Luminous means having the power to reveal, like a light. Without consciousness, nothing can appear. Knowing means having the ability to recognize or apprehend whatever appears. If we experience a sunset while awake or dreaming, consciousness reveals the image and also conceptualizes the image as a setting sun.

In the earliest Upanishads, consciousness is described as having three forms — waking, dreaming, and dreamless sleep. These were originally thought of as places or locations where the inner person traveled — when you dream, your inner self leaves the physical world and travels to the dream world. In later texts, these *places* evolved into *states* or *modes* of consciousness, and four states were identified instead of three: waking, dreaming, dreamless sleep, and pure awareness.

Waking consciousness, sometimes called *gross consciousness*, is concerned with the outer world of perception, and it experiences the self as a physical body, belonging to a person with a life in the exterior world. The world shows up for us through perception, mostly through the visual experience that begins as our eyes collect photons of light.

Dreaming consciousness turns inward and experiences the self as the dream ego, enjoying the images and stories fabricated from past experiences and memories, and probably other deeper sources. In the dream state, the world appears as mental images that we misapprehend as real. Our experience seems real, but only when we wake up from a dream do we know it was a dream. Both waking and dreaming consciousness are restless, dissatisfied, and plagued by desire, as attention constantly jumps from one thing to another.

Deeper still is the consciousness of dreamless sleep. In the Western view, this is *unconsciousness*, the complete absence of consciousness, like that induced by anesthesia. In the Eastern view, dreamless sleep is an unknowing but blissful state resulting from the absence of images, thoughts, and desires. An experienced meditator reaches this state, in which the self is free of desire and the whirling of the mind ceases; consciousness becomes quiescent.

The fourth state of consciousness, underlying wakeful consciousness, dreaming, and dreamless sleep, is pure awareness. This is described as pure non-dual awareness, the supreme wakefulness that reveals the true self as the witnessing awareness behind waking, dreaming, and dreamless sleep. This is the substrate, or ground of being, that underlies all other states of consciousness and is characterized by the quality of *luminosity*.

These four states of consciousness are not four separate things but are contained within each other and interpenetrate within each of us. Waking and dreaming consciousness can exist together in *lucid dreaming*, when you are aware that you are dreaming.

According to scholar of Buddhist studies Jay Garfield, Buddhists think of the human being as being composed of five *skandhas*, or *piles* of phenomena[37] (you do not need to memorize these!):

1. *Rupa* - the physical
2. *Vedana* - the sensory (hedonic)
3. *Samjna* - the perceptual (discriminative)
4. *Samskara* - the dispositional (cognitive and affective)
5. *Vijnana* - the conscious (that which enables knowledge)

As a first approximation, we could say that the five skandhas translate together into our single word, "consciousness." But even within the five skandhas of Buddhism the same 800-pound gorilla lurks: the self. Consciousness is experienced by *someone*. The sunset is revealed to *you*, and *you* apprehend it. Some *self* sees, hears, smells, tastes, feels, thinks, and gains knowledge. Consciousness is nothing without a self to experience it — a movie showing in an empty theater. But what *is* this self, this feeling of *I*

---

[37]. Garfield, Jay L. *Engaging Buddhism*. Oxford University Press (2015).

am, of being the same individual through time who has a unique first-person perspective and a life story?

In Sanskrit the term for the self, *ahamkāra*, means "I-making," which captures the idea that the self is not a *thing* but a *process*. The self is a process of "I-ing," a process that enacts an I, in which the I is no different from the I-ing process. This parallels Damasio's model of the self as an interconnected set of brain processes — *selfing* rather than *a self*.

But Buddhism often seems conflicted on the matter of the self. Hinduism, the precursor of Buddhism, asserts that *Atman* is the *true self*, and it is thought of as the permanent, enduring soul or spirit. However, Buddhism is based on the doctrine of *Anata,* which holds that there is no unchanging, permanent self or soul. Garfield states simply that "in the Buddhist framework … there is no self." This makes sense if you go looking for the self because you won't be able to find it, so on that score we might claim that the self does not exist. Yet centuries of Buddhist scholars have discussed and debated this and concluded that *the self is real and does exist*, and our own personal experience strongly suggests that as well!

A continuing identity is necessary in order to have a sense of autobiography, to connect past experience with the present and our present intentions with future actions. Every morning we wake up and remember who we are and what our life story has been because we have a stable, ongoing self. This is also critical in the legal and moral spheres, where a person who commits a crime is later held responsible for the act because they are still the same self. Another argument Buddhists make for the existence of the self is the subjective unity of consciousness, what psychologists call *binding*. We are able to bind together the many aspects of consciousness, along with ongoing experiences and past memories, into a single subjective representation — *your* experience of being alive, awake, and conscious. The sixth-century Buddhist philosopher Uddyotakara argued that this can only be explained by the presence of a self.

The confusion over self or no-self in Buddhism seems to be a matter of semantics and terminology more than substance. Buddhists speak of a *con-*

ventional or *nominal* (in name only) self that enables us to navigate everyday life and gives us a sense of continuity from one day to the next. But as we navigate life, we can create appearances, play roles, and put on masks as we move through the social milieu — either consciously or unconsciously. This "mainstream" self is often based in fabrications and even delusions, but underlying this is a deeper more authentic self. This perspective, the Buddhist *Middle Way*, avoids both *nihilism* (absolute non-existence of self) and *eternalism* (existence of an absolute self, an ego). Tibetan Buddhists sometimes speak of eradicating the ego, but "ego" does not equal "self." It is the base-level *kind* of self that is prevalent in the world today. As we will see later, it is an evolutionary stage.

The matter of the self also comes up when we ponder death and ask, is there some part of *me* — some part of my *self* — that goes on after my body dies? Western materialist science has nothing to say about death, except *game over*. The answer to this question in Christianity and Islam[38] is that there is a *soul* and an afterlife in heaven or hell, depending on how things went in this life.

But in the Eastern traditions, and in many other religious traditions, the answer to the after-death question is *reincarnation*. In this view there is an essence or soul that goes on after death and finds another body and another life to be born into. In most theories of reincarnation, the soul carries a *karmic record* of a person's life and deeds that will determine the underlying conditions of the next life.

Tibetan Buddhism has far surpassed Western science and religion in its exploration of death. In the *Tibetan Book of the Dead*, a primary text of the Nyingma tradition in Tibetan Buddhism written around 1360 CE, the stages, or *bardos,* of life and death are clearly laid out and explained. When a person is approaching death, they enter the *bardo of dying*. Immediately after death, they enter the *bardo of light*, the clear light or luminosity of pure awareness. This is the ground state of being and does not depend on

---

[38]. According to Rabbi Michael Kosacoff, Judaism is largely agnostic on the after-death question because it is primarily *this-life* focused. The concept of heaven and hell solidified in the fourth century CE Roman Catholic Church and made its way into Islam in the seventh century.

the body and brain which have died and are no longer needed. The *bardo of becoming* is the last interval in the after-death process encompassing the experience of wandering in the form of a subtle *mental body* seeking a new physical embodiment. This stage ends when one enters the womb of one's future mother, and a new cycle of life begins.

Of course, Western science cannot verify the Tibetan account of the afterlife, but the stark reality of death, which completely removes the brain from the picture, crystalizes the age-old question that neuroscientists, philosophers, and even the Dalai Lama still ask. Is consciousness *solely* dependent on the brain, or does some part of consciousness transcend the brain? The modern Buddhist view on this seems clear. The first three states of consciousness — waking, dreaming, and dreamless sleep — all depend on the brain, so when the brain is gone, these types of consciousness are gone. But the fourth state of *pure awareness* remains — it is independent of the brain. If this is true, then consciousness has some foundational platform apart from our brain and our body that would still exist after death.

Consciousness Outside the Brain

While the association of consciousness with the brain is a fairly recent development that arose with modern science, in traditional wisdom no special importance in placed on the brain. Perhaps the oldest and most common view of consciousness is that it is spread out through a system of seven energy centers located from the base of the spine to the top of the head. Hindus use the Sanskrit word *chakra,* which means wheel, for these energy centers. The chakra system has been recognized across many cultures, including the Chinese, Indian, Tibetan, Egyptian, Greek, Judaic, and Islamic, but the earliest known understanding of the chakras comes from Native Americans. For at least 15,000 years the Hopi people have lived on the North American continent (now residing in Arizona) and have long understood that the seven chakras are the seat of consciousness, and they each glow with light to create an aura that emanates from the body. The aura is a reflection of our health — physically, emotionally, mentally, and spiritually.

Each chakra is associated with a particular aspect of consciousness, referred to as a *body*. Rosalyn Bruyere, a world-renowned energy healer, summarizes this in her book, *Wheels of Light* (1989): [39]

2-3
SUMMARY OF THE SEVEN CHAKRAS

	Location/Common Name	Associated Gland	Associated Body	Related to
1st Chakra	Root	Gonads	Physical	Physical sensation, pleasure and pain, survival instincts
2nd Chakra	Sacral	Gut/lymphatic	Emotional	Emotions and feelings
3rd Chakra	Navel	Adrenals	Mental	Intellectual activity, thoughts, opinions, judgements
4th Chakra	Heart	Thymus	Astral	Portal to higher dimensions, bridge between matter and spirit, second feeling
5th Chakra	Throat	Thyroid	Etheric	The "light body" that underlies the physical body, justice and truth, speech and self-expression
6th Chakra	Third Eye	Pineal	Celestial	All forms of seeing: visualization, the future, insight, inspiration
7th Chakra	Crown	Pituitary	Ketheric	The spiritual, mergence with God, Oneness, the All

Mainstream science and Western medicine do not recognize the chakras or the aura that emanates from them because these energies cannot be easily detected with measuring instruments. However, Valerie Hunt (1916-2014), a professor of physiology at UCLA, conducted research on human energy fields for over fifty years and found overwhelming evidence that these energies do exist and are the essence of our consciousness. Hunt conducted years of experiments with subjects wired up to high frequency biofeedback sensors and correlated these data with the descriptions from people able to see auras.[40]

Aura readers like Rosalyn Bruyere see a field of colored light around living things, extending 6-12 inches out from the surface of the body. They see the chakras as wheels of light spinning clockwise in healthy people, and each chakra emits a characteristic color. According to Bruyere, most

---

39. Bruyere, Rosalyn. *Wheels of Light*. Touchstone Books (1989).
40. See Hunt, Valerie. *Infinite Mind: Science of the Human Vibrations of Consciousness*. Malibu Publishing (1995).

"civilized" people have a predominance of yellow light in their aura, emanating from the 3rd chakra because of their intellectual orientation. When someone is unhealthy in some way, the aura shows dark areas, and a skilled energy healer can restore health through energy balancing techniques.

Bruyere recounts that, as a child, her great-grandmother

> …taught me to see auras around plants, although she never used that term. She would say, 'Do you see the light? Do you see that the light on the plant is increasing?' She told me that cuttings taken from a plant on which light was increasing would take root, but that cuttings taken from a plant on which light was decreasing wouldn't root. Then she'd have me do experiments in which I'd take cuttings from different plants, some with a full bright aura, others with a diminishing auric field, so I could discover for myself the truth of her teachings. So it was that in a very practical way I learned to "see" auric light and energy and to identify what was "full of light" as that which also was "full of life."[41]

However, when Bruyere was seven years old, some of the family became concerned about her great-grandmother's mental stability and subjected her to electroshock treatments. After that, there were no more trips to the garden. The seven-year old decided that if she were to avoid this experience, she would stop seeing light around plants, which she did. Then in her twenties she married and had children, and they began to speak about the "colored fuzz" around people. This re-stimulated her own interest and she went on to become a renowned aura reader and energy healer.

Many people in the Western world are skeptical that any of this is real, yet there is nearly universal agreement across the wisdom traditions about the energetic nature of human health and consciousness. This subtle energy has been called prana, chi, qi, mana, life force, orenda, and ruah, among many names. But even in mainstream science there is growing evidence of

---

41. Bruyere, Rosalyn. *Wheels of Light*. Touchstone Books (1989).

an energetic field associated with living things. In 1992 a panel of scientists investigating alternative healing methods at the U.S. National Institutes of Health defined the term *biofield* as a "a massless field, not necessarily electromagnetic, that surrounds and interpenetrates the human body,"[42] later extending this to all living things. Since then, an active field of scientific research has emerged to explore the nature of the biofield. In a 2015 article[43] reviewing the scientific literature on biofields the authors state:

> The properties of such a field could be based on electromagnetic fields, coherent states, biophotons, quantum and quantum-like processes, and ultimately the quantum vacuum. ... The existence of the biofield challenges reductionist approaches and presents its own challenges regarding the origin and source of the biofield, the specific evidence for its existence, its relation to biology, and last but not least, how it may inform an integrated understanding of consciousness and the living universe.

We must conclude that consciousness is not limited to the brain but is spread throughout the chakra system and the entire body. The role of the heart in consciousness has long been understood in the wisdom traditions, but it is largely unknown in Western science. Indigenous people who live close to nature experience life very differently from the way urbanized people see and experience the world. They seem to perceive things that we cannot see. The explanation for this is simple but profound: When you ask them where in the body they live, they gesture to the region of their heart, while modern Westerners typically point to their heads. When we locate consciousness in the brain, we reduce the breadth of full perception and cognition to a narrower band — the brain part of the mind. Heart-centered people perceive things and know things that head-centered people (most of us) do not register.

---

42. Rubik B, et al. *Manual Healing Methods*. From *Alternative Medicine: Expanding Medical Horizons, A Report to the NIH*. U.S. Government Printing Office (1994).
43. Kafatos, Menas C, et al. *Biofield Science: Current Physics Perspectives*. Global Advances in Health and Medicine (November 2015).

Recent research by Rollin McCraty and his team at HeartMath Institute[44] suggests that the heart is like a second brain that is highly connected and synchronized with the cranial brain. The heart can act as a "mind" and an organ of perception because approximately 60 percent of its cells are neurons which function similarly to those in the brain. They cluster in ganglia and form informational networks, and they connect to the neural network of the body. The heart is hard-wired into the amygdala, thalamus, hippocampus, and cortex — brain centers involved with emotional memories, sensory experience, the extraction of meaning from sensory inputs, problem solving, reasoning, and learning. To enhance communication with the brain and central nervous system, the heart also makes and releases its own neurotransmitters as it needs them. The connections to specific centers of the brain create a direct, unmediated flow of information from and to the heart, according to research at the University of Arizona.[45]

Evidence is accumulating that the heart is an organ of perception, intuition, and feeling — a secondary brain that tries to work coherently with the domineering cranial brain. The good news for we civilized people who are trapped in our heads is that we have lots of room to grow in our conscious capacity if we can re-awaken our dormant hearts. This is beginning to happen in the modern world, where brain-centered humans are learning to recognize, open, and strengthen their heart connection; perhaps humans are evolving towards a more fully integrated capacity across the entire heart-brain system, bringing new capabilities.

The heart is mentioned in nearly all ancient texts across all religions and philosophies. It is more than a quaint superstition that the heart is associated with the capacity to love, to know truth, to have intuition, and even to see things before the brain does. The heart is another dancer in the complex interplay of mind and self that is consciousness. We will explore heart intelligence further in Chapter 6.

---

44. McCraty, Rollin. *Science of the Heart, Volume 2*. HeartMath Institute (2015). www.heartmath.org.
45. Song, LZ, Schwartz, GE, Russek, LG. *Heart-focused Attention and Heart-Brain Synchronization: Energetic and Physiological Mechanisms*. Alternative Therapies in Health and Medicine, Vol. 4,5 (1998).

Conclusion

We have taken a whirlwind tour of consciousness and the many attempts over thousands of years to understand it, yet we barely scratched the surface. But "consciousness" is only one word, and we must use it to represent a complex ensemble of interior human capacities. Having some sense of agreement on its meaning will be helpful in the chapters ahead as we explore the *evolution* of consciousness.

To keep us on the same page with a common understanding, let me offer this summary of the meaning of *consciousness* as it will be used from here forward. At the core of consciousness is Antonio Damasio's *self plus mind*, as his book titled *Self Comes to Mind* conveys. I have suggested that the self is at the core of consciousness because consciousness is experienced by a self. The word *mind* has been used in many ways, sometimes as a synonym for consciousness, but I will narrow the definition of mind to be those cognitive functions supported by the cerebral cortex of the brain, including thought, perception, imagination, attention, memory, reasoning, language, and learning. To this we must add in the *emotions* that are generated primarily by the amygdala and brain stem. Some scientists might include the emotions with "mind," but we will view emotions as something organically separate from the cognitive domain that feed upward to the cortical mind, adding rich emotional texture to our mental experiences. Stretching the bounds of science we will also add the *heart* to the ensemble of consciousness, bringing to us the capacity for intuition and love. Finally, to round out the components of "consciousness," we must not forget *the body*, or the wholly embodied experience of personhood and the maelstrom of sensations and feelings that come with that.

These are certainly not separate parts, like the parts of a car engine. Mind, body, emotion, heart, self — this is a system of systems that evolved together, inextricably intertwined. Human consciousness, then, is the beautifully orchestrated ensemble of processes that blend together to give us the felt experience of *being alive and aware*, and *being capable of conscious choice and action*. The self stands at the center, for without it we could not have the *experience* of consciousness — of awareness — or the ability to choose. The word "consciousness" must now serve as a container that holds all of this, all of these interior processes and capacities that humans have acquired. As we now begin to consider the *evolution* of consciousness, we are talking about *all of this* evolving.

CHAPTER 3

Structures of Consciousness

We have been exploring the nature of consciousness in humans, but what about other living things? Do other forms of life have consciousness, equipped with a self and a mind? I feel certain that my faithful dog has consciousness of some kind because she seems to be a distinct being with perceptions, emotions, and thoughts. But she's not very sophisticated in some ways — she likes to roll in decaying carcasses, and she sees me as God-like. It seems clear that dogs are conscious but not in the same way as humans. Most people would say that humans are "smarter" than dogs, but what does that really mean? Dogs are in fact superior to humans in some *capacities*, such as their sense of smell, but their overall *conscious capacity* is different, and — I'll say it — less. They are capable of much less than we are.

The question, then, is not whether some organism has consciousness or not, but what is its *degree* or *level* of consciousness? If we assign some level of consciousness to dogs, then certainly we must do the same for chimps, birds, fish, and reptiles. What about amoebas? Or bacteria? Or trees? Was there a certain point in the evolution of life where consciousness first emerged?

One view, based on the assumption that consciousness arises from neural networks, is that the emergence of the neuron and its electrical connectivity marks the advent of consciousness on Earth. The first multicellular organisms probably evolved electrical signaling — the neuronal cell — for communicating and coordinating throughout their increasingly larger bodies, and the fossil record shows that simple multicellular organisms emerged sometime between 2.0 and 1.5 billion years ago. That's one answer.

Another view, which I prefer, is that consciousness is a property of life itself, and it began on Earth when life began. In the earliest life forms, the cell membrane separated *inside* from *outside*, establishing the simplest kind of identity, a kind of primordial self. Within the cell is *me,* and the outside environment — the world — is something that is *not me*. Even the simplest bacterial cell has a self-interest — the need to survive — while rocks don't. Perhaps consciousness began as a "primordial self," served by a "primordial mind" in the form of chemical sensing and signaling within the cell. Together this primordial self-and-mind could have co-evolved, as chemical and electrical signaling became more sophisticated, and eventually neural networks turned into brains.

From the beginning, in this view, consciousness was evolving along with life itself. Life and consciousness are inseparable, and after almost four billion years of evolution, life has evolved into humans who have the most complex brain and the greatest conscious capacity — the highest level of consciousness — among all of Earth's living things.

Some biologists are reluctant to say that humans are at the pinnacle of evolutionary development here on Earth. Instead, they will say that humans are just another large mammal that came along recently and will probably go extinct, as most species do. Nothing special. Another popular view is that the intelligence of dolphins and whales is on par with humans (and maybe even superior because they are not destroying the planet). I sympathize with this idea, considering that dolphins and whales have sophisticated communication systems using sound, and, who knows, maybe they have deep philosophical discussions or communion with God. But they apparently produce nothing that lasts beyond a lifetime, and their impact on the world is minimal. We see no evidence that dolphins have created poetry or art or any technology, or have any kind of evolving culture. Humans are in another league. No other living thing has acquired the brain, the mind, the self, the heart, and the body — all working together as a whole to produce the level of consciousness that we have. I'm not saying we're *better* than whales and chimps, but we must own up to, and take responsibility for, the awesome power we have.

It is best to think of consciousness on an evolutionary continuum, with bacteria at one end and humans on the other. As life has evolved into greater complexity, consciousness has also expanded and reorganized into new regimes that support higher functioning and greater capabilities. There is a clear correlation between *capability* — what and how much an organism can *do* in the world — and *conscious capacity*. With greater conscious capacity we are capable of more. Crocodiles are more capable than fish, but dogs are more capable than crocodiles, and humans are more capable than dogs, all because of greater conscious capacity.

Humans certainly have demonstrated capabilities and powers far beyond any other living thing, so, yes, I am making a case that humans do indeed sit at the pinnacle of evolution here on Earth. But it is the evolution of consciousness that has made humans the overwhelming force that now dominates the biosphere. Without our evolved consciousness we *would* be just another large mammal, but humanity is a form of life unlike any other, capable of causing a global extinction event, as well as creating art and science.

The unparalleled brain growth that occurred in the *Homo* lineage over several million years finally stopped with the appearance of *Homo neanderthalensis* around 400,000 years ago and *Homo sapiens* by about 300,000 years ago. The physical brain reached full size long ago, yet *culture* has continued to evolve, spectacularly, right up until today. We can follow the evolution of culture in the archaeological record (mostly stone tools) and see that human capabilities have expanded continually over the last few million years, with dramatic transformational periods like the Great Leap about 60,000 years ago and the emergence of civilization 5,000 years ago. Underlying this evolution of culture — what made it happen — was the *evolution of consciousness*. Something was going on *inside* that skull and brain: new structures, new connections, and new modes of cognition, altogether comprising the evolution of consciousness. This is what separates humans from all other living things.

The idea that consciousness evolves took hold in the West in the early 1600s and has been explored continually by a long and venerable line of philosophers, including Leibniz, Kant, Goethe, Fichte, Hegel, Schopenhauer, Spencer, Bergson, Whitehead, Aurobindo, and Teilhard de Chardin. In the late 1800s, the field of psychology was coming into its own, having diverged from philosophy, and in the early 1900s developmental psychology was established, largely through the work of James Mark Baldwin. Baldwin and the developmentalists who followed have proposed stages of human development and have identified universal, cross-cultural stages of child and human development across multiple domains, including cognition, morality, and the self. These pioneering developmentalists include Sigmund Freud, Carl Jung, Jean Piaget, Abraham Maslow, Clare Graves, Erik Erikson, Jane Loevinger, Lawrence Kohlberg, Robert Kegan, and most recently, Ken Wilber and Don Beck.

However, there was one scholar in Europe who was decades ahead of his time and discovered the foundational template for the structure and evolution of consciousness. Jean Gebser (1905 — 1973) first published his ideas in 1949, yet his work has remained largely unknown outside of Europe until recently. Just as any discussion about the evolution of life starts with Darwin, our quest to understand the evolution of consciousness must begin with Gebser.

The Ever-Present Origin: Jean Gebser

Jean Gebser in 1957
Public Domain

Jean Gebser was born in Posen, Imperial Germany into a family that was affluent and well-educated. But his father committed suicide when he was seventeen, plunging the family into poverty. Gebser was forced to abandon school and became an apprentice at a bank in Berlin. The drudgery and boredom were tolerable only because in his spare time he attended lectures at Berlin University, where he discovered the work of Rilke, Schopenhauer, and Freud. In 1929 Gebser decided to leave Germany after witnessing the first "brown shirts" of the Nazis. He lived in Spain for the next six years, where he befriended and worked with Federico Garcia Lorca and other poets, whose works he translated into German.

Twelve hours before his apartment in Madrid was bombed in the fall of 1936, Gebser abandoned everything and fled to Paris where many other intellectuals were seeking refuge. There he shared company with a circle that included Pablo Picasso, Andre Malraux, Paul Eluard, and Louis Aragon. By 1939, World War II had erupted, and Gebser decided to leave France. Two hours before neutral Switzerland closed its borders, he crossed into safety. Gebser finally found a permanent home in Switzerland.

In the following decades, he worked tirelessly to give shape to the inner vision that first appeared to him in the winter of 1931 when he was 26 years old. He realized that the spectacular developments in the arts and sciences during the first three decades of the twentieth century were the beginning of a transformation in the consciousness of humanity, in the way we experience ourselves and the way we see the world. Gebser compared the significance of this transformation to the Axial Age, which took place 2,500 years ago in the time of Socrates in Greece, Lao-Tzu in China, and the Buddha in India.

During his time in Switzerland, Gebser socialized and worked with the likes of Werner Heisenberg, Lama Govinda, and Carl Jung, and he visited D.T. Suzuki in Japan. He was a frequent lecturer at the Institute for Applied Psychology in Zurich, and a member of Jung's Eranos circle that included Adolf Portmann, Mircea Eliade, Karoly Kerenyi, and others at the leading edge of culture and thought. Gebser became highly regarded throughout Europe, where he lectured widely, received numerous literature awards, and was appointed to an honorary professorship at the University of Saltzberg six years before he died in 1973. Yet today, his name and his work are just being discovered more widely in North America.

Gebser is often called a cultural philosopher, but his writings encompass anthropology, philosophy, sociology, psychology, linguistics, mythology, poetry, history, theology, Indian spirituality, and many fields of science. His work has been difficult for traditional scholars to access because it does not fall into conventional categories, and he worked largely outside of academia. Gebser published many books, ranging from translations of poetry to commentaries on social and cultural evolution, but his magnum opus was *The Ever-Present Origin*, originally published in two parts in 1949 and 1953. It was not available in English until 1985.

In Part I of this 500-page tome, Gebser proposes and details four fundamental structures of consciousness that are within humans today and he tracks their evolutionary emergence based on the best knowledge of his time in cultural anthropology, world history, literature, and mythology. Part II is devoted to describing a fifth and new structure of consciousness that is just beginning to emerge in humanity. His several names for this new consciousness include the *aperspectival*, the *arational*, and, most often, the *integral consciousness*. As far as we can tell, Gebser was the originator of the term *integral consciousness* that is widely used today. It first appears in his writings in the winter of 1931.

The Five Structures

The Ever-Present Origin takes on the primary questions posed at the beginning this book: what is the phenomenon of humanity, what is our story, and what is our future? Answering these in 1949, Gebser begins by defining five ways of seeing the world and being in the world, which emerged successively as our ancestors evolved. These are initially called the *structures of consciousness,* and Gebser names them:

1. The Archaic
2. The Magical
3. The Mythical
4. The Mental
5. The Integral

As structures of consciousness we can think of these as worldviews, or reality sets, that evolved over time, one on top of another, since at least the time of *Homo erectus*. How we experience reality is called *phenomenology* by philosophers, and Gebser understood that the evolution of consciousness was fundamentally phenomenological — our experience of reality was changing. Because these structures, these ways of experiencing the world, evolved over time, they also correspond to *stages of evolutio*n. They describe the evolution of consciousness. Furthermore, because of the close relationship between consciousness and culture, these are also *stages of culture* that show up in the archaeological record.

However, in the 1940s when Gebser was working on this, scientists knew almost nothing about the deep history of our lineage — the field of paleoanthropology scarcely existed because only a few handfuls of ancient human specimens had been discovered (and most of those were *Homo erectus* specimens from China, so-called Peking Man). In those days the great debate was about whether humans originated in Asia or Africa, and Asia was the majority opinion. This was finally settled in 1960 by Louis and Mary Leakey's discovery of *Homo habilis* in Kenya, and since then, the field of paleoanthropology has exploded.

As we now explore Gebser's work, we have access to a prolific new field of science, paleoanthropology, that has exploded our knowledge of human evolution. Researchers have carefully uncovered and studied thousands of skeletal specimens and artifacts from all over the world, and they now use technologies like radio-isotopic dating and genome mapping. Today paleoanthropologists have extensive knowledge of the last 3 million years of human evolution, as summarized in Chapter One, even though many mysteries remain. As we explore Gebser's five structures in the evolution of consciousness, I will supplement his work with the most current findings from paleoanthropology.

Considering that Gebser had virtually no evidence or knowledge of human activity before about 5,000 years ago, how did he arrive at these evolutionary structures and stages that we now know span 3 million years? What Gebser *did* have was a deep knowledge of the humanities — literature, mythology, art and architecture, world religions, and, importantly, *cultural* anthropology. In the first half of the twentieth century, cultural anthropologists, like Margaret Meade and Gregory Bateson, were able to study the vanishing indigenous societies that were still insulated from the Western "civilized" world, in places like Samoa and New Guinea. These cultures are now largely gone, as human population has exploded and civilization has spread everywhere, but Gebser was able to draw great insight from these studies.

Gebser used this knowledge to work backwards from our present era and mindset, the Mental stage and structure of consciousness. He pinpoints the origin of the "modern mind" to classical Greece about 2,500 years ago in the so-called *Axial Age*, a widely held view. Having first emerged in Greece, the Mental consciousness reached its pinnacle with the European Renaissance and the development of modern science, but it also began declining and stalling out in the twentieth century with the World Wars and the atomic bomb.

Of course, Gebser understood the Mental consciousness very well, as we all do, because it is still our current operating system. Our base reality set is a three-dimensional world of objects in space, with a sense of linear time that flows from past to future. We remember the past and have a life story (or autobiography), we live in the present, and we imagine the future. This is the nature of the Mental structure of consciousness, which is the dominant, or governing, structure for modern humans. But Gebser knew that something came before this, before classical Greece. From his knowledge of Homeric Greece and the mythologies and the oral traditions of the world pre-dating writing, he fleshed out an earlier era of human evolution, the *Mythical* consciousness. This was a dreamy, tribal reality in which the spoken story was the highest level of meaning.

Working backwards even further, Gebser surmised a pre-Mythical consciousness based on the known cultural anthropology of primitive societies. He called this the *Magical* consciousness and considered it to be earliest stage of distinctly human consciousness, at the dawn of human evolution. Of course, he was completely unable to date the Magical era or know what was before this as our ancestors diverged from the apes. The pre-Magical consciousness was unknown and mysterious, a sort of core consciousness bridging the animal and the human domains. He named this the *Archaic* consciousness, the first of his structures.

These four structures become a framework for the story of humanity; our ancestors progressed through these four cultural stages, reflecting the evolution of consciousness from the Archaic to the Magical to the Mythical, and finally to the Mental structure where we still operate today. Gebser's stages and structures are now widely recognized by modern scholars of culture and consciousness, including Ken Wilber, Steve Macintosh, William Irwin Thompson, Gary Lachman, and Don Beck.

Arguably, Gebser's most inspiring contribution was his vision of the future. In 1931, he realized that the development of relativity and quantum physics, and other cultural transformations in the early twentieth century, were signs of an emerging new consciousness in humans. In his writings from this time, he calls this the *integral* consciousness, and he spent the next forty years writing and teaching about the five structures and the integral consciousness. He saw the integral consciousness as a very hopeful and positive vision of the future of humanity.

What, then, is the *meaning* of the five structures — the Archaic, the Magical, the Mythical, the Mental, and the Integral? How should we think of them? Let's summarize this in four ways:

1. As *structures* these are the architecture, or the landscape, of human consciousness. They evolved successively, each one building on the previous. Each structure is a different worldview or reality set, but earlier structures are always retained. Each structure *transcends but includes* the previous, as depicted in 3-1. We can recognize this as the fundamental structure of nature: an atom transcends and includes its constituent sub-atomic particles, a molecule transcends and includes its atoms, a cell transcends and includes its molecules, an organ transcends and includes its cells, and a body transcends and includes its organs.

2. Being human ourselves, these are the structures of *our own* consciousness. Therefore, Figure 3-1 is also a representation of our own consciousness. Stories are meaningful to us because we still have a Mythical consciousness; experiences in nature activate our Magical sense; but our Mental structure overrides these with its excessive thinking and the need for reasoned explanations.

3-1
GEBSER'S STRUCTURES OF CONSCIOUSNESS

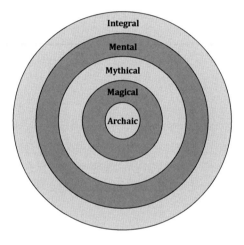

3. Because these structures evolved in succession, these are also *stages of evolution*, periods of history when a certain structure of consciousness governed and expressed itself as a stage of culture. These are evolutionary stages of *both* culture *and* consciousness. Gebser could only put these stages in order, but he could not assign any dates, except to associate the beginning of the Mental stage with fifth century BCE Greece.

4. Gebser proposed another way of understanding the nature of these structures. He recognized a parallel between the evolutionary stages of our lineage and the life stages of individual humans, as proposed by Jean Piaget, Abraham Maslow, and others. For example, Piaget found that the growing child is not capable of abstract thinking until about age 12; that is, they do not acquire the Mental structure

of consciousness in their own cognitive architecture until this time. In this sense, our life is a reenactment of the ancestral stages in the evolution of consciousness. The newborn comes into the world with only the Archaic consciousness and later acquires each new structure as they grow and progress through life. This begs the question, can adults who have achieved the Mental structure (nearly everyone today) move beyond this and into the Integral structure? The remainder of this book addresses this question.

Let's now explore the nature of Gebser's five structures in more depth.

The Archaic Structure

In the Archaic consciousness, there is no awareness of space or time as we know it, and there is no sense of self as something separate from the environment. Subjective and objective are one. According to Gebser, the archaic structure is zero dimensional (spaceless) and timeless, and there is "complete non-differentiation of man and the universe." In this instinctual mode the concern is only with the present moment. The Archaic consciousness has been likened to deep, dreamless sleep and, as mentioned earlier, can be understood by considering the consciousness of a newborn child. Georg Feuerstein, probably the foremost interpreter of Gebser, describes the archaic consciousness of the newborn:

> At the neonatal stage, consciousness is still undifferentiated. It lacks a defined subjective center, and hence there are no objects for the newborn. The neonate is one with the sensations that flood through the body. Only at the age of one month does a baby focus on an attractive object, though he or she will as yet make no concerted attempt to grasp for it. First fumbling attempts at reaching for objects occur toward the end of the second month, and accurate grasping becomes possible only around the fifth month. But even then no real self-consciousness exists in the suckling.[46]

---

[46]. Feuerstein, Georg. *Structures of Consciousness: The Genius of Jean Gebser.* Integral Publishing (1987).

> ... when a newborn awakens to the world outside the womb it experiences its surroundings synesthetically. That is to say, it smells sounds and hears forms: its sensations are all wired together, and only as the nervous system develops and comes under the spell of the left cerebral hemisphere does this capacity vanish.[47]

From the perspective of paleoanthropology, we should think of the Archaic structure as characteristic of pre-*Homo* species, such as *Australopithecus africanus* or *Homo habilis*, but in Gebser's time there was no knowledge of these species.

The Magical Structure

In the early Magical stage, our ancestors began to experience a separation from nature, in the same way a one-year old must begin to separate from his or her mother. Jeremy Johnson writes:

> The boundaries between self and world, culture and nature in this structure are more like a permeable cell wall than a city wall, however, and so the consciousness of the magical is marked by the capacity to *merge* with nature, to slip into trance ...[48]

According to William Irwin Thompson, in the Magical structure:

> The emerging personal ego is not yet stabilized and the labile mind can function in both the dreaming and waking consciousness at the same time.[49]

In the Magical structure, Gebser says, "every*thing* intertwines and is interchangeable."[50]

---

[47]. Feuerstein, Georg. *What Color is Your Consciousness?* Robert Briggs Associates (1989).
[48]. Johnson, Jeremy. *Seeing Through the World*. Revelore Press (2019).
[49]. Thompson, William Irwin. *Beyond Religion*. Steiner Books (2013).
[50]. All Gebser quotes are from *The Ever-Present Origin*, Ohio University Press (1986).

Johnson elaborates:

> Any one point is a secret passage to all points. Yesterday, today, and tomorrow are all interchangeable. Ritual acts are inseparable from their desired outcomes, and they are not causal as we would think (that X ritual produces Y result). Instead, for the magical structure, X is Y.[51]

At about the age of one, the child begins to realize they are separate from their mother and alone in a big, scary world. The journey into selfhood begins in the magical world of the toddler who delights in being discovered in a game of peek-a-boo. But at the same time, anxiety and a sense of aloneness is growing. Separation from mother is a central issue for children of this age, and, so too, our ancestors had to confront their aloneness in the world.

Just as the one-year old must deal with fears and nightmares, people governed by Magical consciousness experience nature and its forces with fear and awe. Offerings had to be made to appease the nature-forces, and emotion-filled shamanistic rituals with chanting and dancing and drumming were essential for dealing with a world filled with nature spirits. People living in Magical consciousness banded together in clans, with an identity that was "we," but not yet "I." For the toddler, the extended family, growing beyond mother, father, and siblings, is the clan.

3-2
Wandjina rock art from Western Australia.
Keith Michael Taylor/Shutterstock

Gebser also explains that the art of Magical people often depicts mouthlessness. He cites the example of the rock art of aboriginal Australians (3-2).

---

51. Johnson, Jeremy. *Seeing Through the World*. Revelore Press (2019).

What this mouthlessness means is immediately apparent when one realizes to what extent these paintings and statuettes are an expression of the magic structure - but not yet of the mythical structure. Only when myth appears does the mouth, to utter it, appear... (It) indicates to what extent magic man placed significance on what he heard, that is, the sounds of nature, and not on what was spoken.

It is not yet known which of our ancestors first moved into Magical consciousness, but I will propose that *Homo heidelbergensis* is a likely suspect, perhaps beginning around 1 million years ago.[52]

It is clear that by about 300,000 years ago, the brain had grown to full size in several species: *Homo sapiens* in Africa, *Homo neandertalensis* in Eurasia (with a slightly larger brain), and the newly discovered Denisovans[53] in Asia. It is widely accepted that all these species were descended from *Homo heidelbergensis*. Artifacts from humans and Neanderthals dating to around 300,000 years ago are remarkably similar, suggesting that the two species developed in a parallel fashion — but on different continents! It seems plausible that Neanderthals, Denisovans, and early humans all inherited the Magical consciousness from *heidelbergensis*.

The Mythical Structure

Gebser made a strong association between the Mythical consciousness and the acquisition of speech and complex language in humans. Some anthropologists today suggest that the precursors of spoken language were emerging several hundred thousand years ago, in both Neanderthals and humans, but sophisticated language with large numbers of words probably emerged *only* in humans around 60,000 years ago, coinciding with the Great Leap. Spoken language required anatomical developments to support vocalizing and speech, which the Neanderthals probably had, and also

---

52. To date, the oldest *heidelbergensis* skeletal remains are about 700,000 years old, but artifacts from Africa suggest that this species was present by at least 1 million years ago and perhaps as early as 1.3 million years ago. I will use 1 million years ago as a reasonable date for the appearance of *heidelbergensis* in Africa.

53. A species name has not yet been given at this writing but some are calling this cousin of ours *Homo denisovensis*.

significant advancements in the neo-cortex that supported the use of words as symbols with meaning. Apparently, this cortical development passed over the Neanderthals. The complex speech and cognitive sophistication acquired by humans amplified their ability to communicate, to teach and learn, to problem-solve, and to pass along knowledge and culture. This catapulted humans past the Neanderthals and all other rivals, giving them major advantages that allowed them to move out of Africa and inhabit the entire planet.

Gebser derived much of his thinking about Mythical consciousness from literature, poetry, and mythology — especially that of Homeric Greece representing the very late Mythical period. Spoken stories and the oral traditions were the source of meaning for people in the Mythical consciousness. Mythical stories with fantastical characters helped explain the world, providing social structure, comfort, and morality. The distinct separation between dreaming and waking was probably not as clear as it is for us, as myth and "real life" blurred together in a dreamy landscape.

The reality of Mythical people was a drama with heroes, villains, and a storyline. People had clear roles to play within the tribe, as though in a Shakespearian play. This was a new type of selfhood — not a fully independent and autonomous person, but an emerging tribal self that was cast in a definite role within the larger social structure. The story-like reality of Mythical people also contained a new sense of time with a forward-moving flow of events — the plot, if you like. All these things still play an important part in our own reality because Mythical consciousness is still very much alive within us. This was famously expressed by Jacques in William Shakespeare's *As You Like It*:

> "All the world's a stage,
> And all the men and women merely players;
> They have their exits and their entrances;
> And one man in his time plays many parts …"

According to Gebser, Mythical consciousness is dream-like and *pre-causal*. As we all know, strange things happen in dreams that could never happen when we are awake. We can suddenly be in a different place, or someone can appear out of nowhere. Our waking world is *causal*, where things happen for a reason and can be explained, but dreams do not adhere to causality, nor does the reality of Mythical consciousness. The understanding of causality emerged later with the Mental structure, reaching full expression in classical physics. For all of us today, causality is simply how the world works, according to "common sense." But causality had not yet been realized in the earlier cultures governed by the Mythical structure.

In acquiring Mythical consciousness, humans made the full separation from nature, and for the first time, people came to understand their own demise: death. This arises from the emergence of the *autobiographical self*—the experience of being a person with a life story, who was born and will also die. Death, the soul, destiny, and the afterlife were all of major concern in Mythical cultures. In the late Mythical stage, the entire civilization of Egypt revolved around the afterlife, and the mythologies of Homeric Greece convey the same concerns.

Gebser discusses at length the changing conception of time and space as consciousness evolves. Magical consciousness is pre-temporal, with only an awareness of day and night, sunrise and sunset (solar time), but in Mythical consciousness there is a growing understanding of time moving forward in cycles (calendric time), based on the phases of the Moon and the passing seasons. Days, months, and years were understood and carefully monitored through astronomical observatories like those found at Stonehenge, Machu Picchu, Angkor Wat, and many other sites around the world. Mythical people could track the solstices and equinoxes, and even predict eclipses.

As human consciousness evolved, so did our conception of space — it was growing in dimensionality. Gebser describes Archaic consciousness as zero-dimensional, Magical consciousness as one-dimensional, and Mythical consciousness as two-dimensional. Mythical people had a flat-Earth cosmology — the very sensible belief that the world we live in is flat. They had no conception of three-dimensional or spherical space, which first shows up in Greece during the fifth century BCE. The Greeks understood that the Earth is round (they knew its circumference), and they even knew the distance to the Moon. Such a comprehension of space was not possible within Mythical consciousness, constrained to two-dimensions.

The early Mythical structure corresponds in child development to the infamous "terrible twos," with temper tantrums and defiance as the individual self begins to separate and coalesce. Limits are tested, and personal power is explored, but as the child ventures out into the world, natural barriers and hazards are encountered: the world is a big and dangerous place. The two-year old is not yet ready to go out in the world as a fully autonomous self, but still must occupy a secure role within the expanding social circle. Just as good parents must set up well-defined boundaries that protect a child without stifling their openness and exploratory drive, so too did the elders of paleolithic tribes establish regulatory strictures, boundaries, and taboos — all conveyed through myth — that helped create order in a chaotic and frightening world. These were the precursors of the laws and morality that would later underpin civilization.

The humans who survived the Toba bottleneck and experienced the Great Leap 60,000 years ago were able to leave Africa easily (unlike their ancestors), spread throughout the habitable continents of Earth, and become the dominating force we still are today. By about 45,000 years ago, modern humans had moved into Australia, Asia, and Eurasia, and by about 40,000 years ago they had moved into Europe to become the *Cro-Magnon* culture. All these people left signs of their mythical culture and expanded conscious capacity as they spread over the entire planet. This is when humans began to acquire the power to do serious damage to the environment, and it is when we bypassed natural selection for all practical purposes. Humans had no peers in the animal world, and with clothing, shelters, and sophisticated tools, humans could live in almost any environment, including Siberia. The Great Leap that transformed humans into a geologic force was nothing more, and nothing less, than the acquisition of Mythical consciousness. The Great Leap was a leap in consciousness.

The Neanderthals, who lived throughout Europe and Eurasia by at least 300,000 years ago, must have fully attained Magical consciousness, and up until about 100,000 years ago, they were at least on par with *Homo sapiens* in their abilities. There is anatomic and genetic evidence that Neanderthals acquired the ability to vocalize and may have used simple spoken language, but apparently, they did not experience the cognitive leap in brain function that provided the support for complex language. We must conclude that Neanderthals did not make the transition into Mythical consciousness and the full acquisition of spoken language, as *Homo sapiens* did. Coming out of the Great Leap, humans had entered a new regime of mind and self — of consciousness — while the Neanderthals plateaued in their development, disappearing not long after the new humans arrived in their homeland of Europe about 40,000 years ago. It was the evolution of consciousness that separated humans from Neanderthals and made humans more powerful and dominant.

In Gebser's theory, after one stage of consciousness becomes dominant, it eventually reaches a point of "deficiency," when it can no longer serve effectively as the highest governing structure. Such a deficiency in the Mythical structure was starting to grow as the first civilizations were emerging about 5,000 years ago. The plethora of gods and contradictory stories of creation and morality were overwhelming. Humans were drowning in a deluge of mythological confusion that no longer helped them understand and function in the world. In reaction to this, Socrates, Plato, Aristotle, Pythagoras, and company, ushered in a new structure and stage of culture that emerged from the growing deficiency in the Mythical structure. According to Gebser, the Classical Age of Greece, in the fifth century BCE, was underpinned by the emergence of the Mental structure of consciousness as the Mythical structure began losing dominance.

The Mental Structure

Gebser explains that the Mental structure of consciousness reached its fullest and most dominant expression in the European Renaissance and Enlightenment, but it first emerged about two thousand years earlier in Classical Greece. Some have called this prolific flowering of thought and culture the birth of the "modern mind." All of us in the modern world today operate predominantly in Mental consciousness, just as Euripides did in Athens 2,500 years ago — we fully identify with the characters in a Greek tragedy as people very much like us.

The cultural explosion that humans experienced 2,500 years ago in Greece also occurred fairly simultaneously in China, India, and Persia. This period has been called the *Axial Age,* a term coined by philosopher Karl Jaspers. New capabilities of the mind emerged as Mental consciousness took form and became dominant: logic, reason, abstraction, extensive written language, mathematics, accelerated learning, and an ever-growing knowledge base. We can suspect that the neocortex was again reorganizing in new ways to support this cognitive leap in conscious capacity. Written language, which first appeared about 5,000 years ago in the early civilizations, had developed into a powerful tool of the mind by the time Euripides was writing tragedies in Classical Greece.

But the great flowering of Mental consciousness in Greece did not simply spread and proliferate. Greece was absorbed by the Roman Empire, and Rome eventually became corrupt and was overrun by the Goths. The knowledge and culture of Classical Greece was lost in medieval Europe during the so-called Dark Ages when the Earth was once again flat and people retreated back to the tribalism of Mythical consciousness. Retreats into past consciousness structures — the devolutionary slide — are always possible because we maintain the previous structures of consciousness.

Yet the flame of Mental consciousness, of Greece and the Axial Age, was kept alight in Constantinople, China, India, and the Islamic world, where mathematics, science, art, and literature thrived and proliferated, while medieval Europe languished in the dreamy feudalism of Mythical consciousness for a thousand years. Finally, in the fourteenth century, Europe awakened in the Renaissance, signaling the re-emergence of the Mental consciousness, which became the dominant paradigm of reality in the Western world, as it remains today.

Two thousand years after the Axial Age, the Mental structure of consciousness found its penultimate expression in the European Enlightenment with the emergence of empirical science. Gebser often referred to this as the *mental rational* or simply the *rational* consciousness, but cautioned that rationality was also limiting. In the 1600s Galileo and Newton showed that nature followed mathematical equations and thus could be predicted and controlled. They had discovered the *mechanical universe*, where causality ruled. This approach that we now call classical or Newtonian physics was highly successful, and it brought ever-greater power to humans. From this came the steam engines that powered the factories and the locomotives and the ships of the industrial revolution, while the telegraph brought nearly instantaneous communication across continents. Europe and the Western world were radically transformed, and by about 1880, science and industry were all-powerful and widely thought to be infallible. This was questioned by the writers and artists of the Romance movement, such as Blake, Rousseau, Emerson, Cole, and Thoreau, who pushed back against the Newtonian machine. But they were unable to slow the mighty freight train of progress.

In the late period of the Mental consciousness, rationality, reasoning, and abstract thinking were held with the highest regard, and all these converged in the development of empirical, or evidence-based, science around 1600. Empirical science became, and remains, the most powerful tool humans have ever developed. This method, this process we call science, has been so effective at discovering the underlying truths of the physical world that our knowledge base has grown faster than our wisdom, and our new technologies have become exponentially more powerful and sometimes lethal. Nuclear weapons and other weapons of mass killing are obvious existential threats, while less visible technologies like gene editing and artificial intelligence could have massive unintended consequences.

Some might argue that the Mental consciousness, and the science it created, has taken us to the brink of destruction, and perhaps even extinction. But I would counter that science and technology are only tools, and like all tools, they are only as good or bad as how we choose to use them. However, modern humans living in the Mental consciousness are apparently not yet able to make wise choices about how to use our powerful technologies. This illustrates Gebser's "deficiency" in any stage, when a governing structure of consciousness is no longer serving us, and things clearly start to go wrong. Gebser was painfully aware of the deficiency that had emerged in the Mental consciousness, having lived through the horrors of two World Wars and the unleashing of the atomic bombs at Hiroshima and Nagasaki.

In Mental consciousness, causality (cause and effect) becomes the source of meaning (things happen for a reason), laying the foundation for the explanations of science. Cognitive psychologist Merlin Donald calls this the *theoretic* mode of consciousness. In science we seek explanations — or theories — for everything, and so do non-scientists. In the theoretic mind, magic and myth are surpassed by the power of logic and rationality. The ever-growing knowledge base could be preserved through external media, such as writing (and, now, electronic media), and passed on to future generations. With the emergence of Mental consciousness, the world became predictable, and with that came a feeling that all of nature could be mastered. The Magical spirits and Mythical gods no longer had to be appeased. The rational mind could slay the unknown and give humans the power to control nature. Of course, this is pure hubris.

In Mental consciousness the linear conception of time emerges and three-dimensional space becomes fully realized. Humans could now comprehend the arrow of time pointing from past to present to future, creating history, autobiography, and family lineages. The Mythical sense of time is cyclic but lacks a forward progression; however, the invention of clocks, starting with the sundial, made time measurable and ongoing. Likewise, the flat-Earth conception of space in the Mythical world was expanded by the Greeks who knew Earth was round.

However, Gebser points out that the artists' technique of creating the illusion of three-dimensional space on a flat surface did not appear until the late 1400s when Brunelleschi invented *linear perspective.* This development demonstrates the full conceptualization of three-dimensional space, transcending and including the two-dimensional cosmos of the Mythical. The combination of three-dimensional space and linear time, governed by causality, creates the exterior phenomenal reality we know so well — a three-dimensional world of independent objects separated in space, interacting according to the laws of cause and effect. People before this who had not acquired Mental consciousness saw and experienced the world in a much different way, much like our dreams.

According to Gebser, the Mental structure is *perspectival* and, necessarily, *egoic*. In the Mental consciousness the self fully individuates and solidifies as something separate from the rest of the world. Space becomes fully realized as *that which separates*. We now look *upon* nature instead of *participating in* it, and we look outward from the perspective of "me." Having a particular perspective, a *single perspective*, requires an *I*, a *me*, a *mine*, and thus is born the modern egoic self.

In the *un*-perspectival consciousness of the Mythical and Magical, space *encloses*, in what Jeremy Johnson calls the *world-as-cave*. The awakening of the Mental consciousness brought us out of the cave and into the daylight of three-dimensional space, extending forever outward and filled with the hard objects of the world. With our eyes we look out at the world from the single viewpoint of *me*, and the dualism of Descartes now emerges — the *subjective* and the *objective*, *mind* and *matter*, the *inner* and the *outer*, *me* and *everything else*.

With the unfolding of each new structure of consciousness, the brain was evolving in its capacity to handle new tasks and support new capabilities, giving humans increasing power over the environment and other living things. But as the power of humans has grown, so has a toxic disconnect with nature. In the Mental structure, humans have been at war with nature, seeking to conquer, to exploit, and to master her. This was at the heart of the European colonization of the world and the brutal subjugation of indigenous peoples. The Aztecs in Mexico, who welcomed the Spanish conquistador Hernan Cortes and his men in 1519, were firmly ensconced in Mythical consciousness and were easily misled by the self-serving lies of the Europeans who were operating in Mental consciousness; the Aztecs were methodically slaughtered as Cortes claimed Mexico for Spain. Cortes and his men were capable of much more — including treachery, deception, and betrayal — than the naïve Aztecs. But this was only the beginning of an increasingly destructive path for humans operating in Mental consciousness, a path that today is leading us over a cliff. Gebser understood this all too well.

The Integral Structure

Gebser devotes Part II of *The Ever-Present Origin* to describing and giving examples of the *Integral* consciousness. He recognized that in the twentieth century the Mental structure had become fully deficient — we might say in today's parlance, unsustainable — opening the way for a new *intensified* structure of consciousness.

Integral means *composed of parts that together constitute a whole*. In this sense, Integral consciousness brings together the first four structures of consciousness into a whole that is something completely new. While Mental consciousness separates everything into parts and categories, Integral consciousness unifies separate parts into wholeness. In Mental consciousness the world and the human being are fragmented; in the Integral they become whole.

According to Feuerstein, describing the Integral:

> It transcends the obsessive tendency of the rational consciousness to think of everything, including aspatial time, in spatial terms only. In the end it transcends the hardened ego itself, which is the fixed point, or pivot, of the rational consciousness. The intensified consciousness, which is attuned to the flow of reality itself, transcends all perspectival distortions of the rational egoic consciousness. … the integral consciousness is about perceiving and communicating that which is real, not merely imagined or conceived.[54]

The Magical and Mythical structures are *unperspectival*, the Mental is *perspectival*, and Integral consciousness is *aperspectival*. The "*a*" means *without*, or to be free of. The Integral is *without* a single perspective while the Mental consciousness can *only* have a single perspective. We are all so fa-

---

54. Feuerstein, Georg. *What Color is Your Consciousness?* Robert Briggs Associates (1989).

miliar with this, it's easy to overlook because the single perspective we are talking about is *one's own personal* perspective. What other perspective can I have? I can only see through *my* eyes, hear with *my* ears, think *my* thoughts, and have *my* opinions, right? Yes, in the Mental consciousness.

But in the *aperspectival* consciousness of the Integral, one leaves the self-centered, single perspective and can simultaneously hold multiple perspectives and change perspectives. This is only possible if the nature of the self evolves. It is the self-centered ego of Mental consciousness that creates the single perspective, but when the egoic self softens and becomes transparent — when you no longer feel you are the center of the universe — the aperspectival world emerges. We will explore these ideas in more depth in Chapter 6.

In the Mental structure, three-dimensional space is fully realized, or *concretized* in Gebser's language. Instead of space enclosing us in the cave-like world of Mythical consciousness, Mental consciousness experiences the infinite emptiness of the universe, making Earth and humanity miniscule. We have tried to conquer space with our technologies, like cars, airplanes, and spacecraft, and we try to control space through ownership, territory, nations, and empires. The remaining challenge for humans in the Mental regime is *time*. Time remains our enemy.

In the Integral we master time by achieving *time freedom*, where past, present, and future lose their meaning and power. In the Mental structure, time is linear, and the present separates the past from the future; the future is unknown and inaccessible, and the past is set in stone. But in the Integral, time becomes whole instead of fragmented, and it is intimately connected to space. According to Gebser, reality becomes four-dimensional. Einstein expressed this mathematically in relativity theory as four-dimensional space-time, and indeed, Gebser recognized Einstein's work as one the first signs of the emerging Integral structure of consciousness.

Linear time and three-dimensional space also underlie causality, or cause and effect, one of the most basic aspects of our world of Mental consciousness. Causality implies the passing of time: if I do *this*, then *that* will happen (later). But with time freedom, where past, present, and future are no longer distinct, causality is not necessary. Gebser describes the Integral as *acausal*, or free of causality. I must confess, I have a hard time picturing that!

The Integral structure is a new way of seeing the world and knowing what's true. In the Mythical structure meaning and knowledge are conveyed through *mythology* — stories about what is true. In the Mental structure we have *philosophy* — *thinking* about truth and using reason and logic to lay out sequential arguments that demonstrate what is true. Science arises from this. But in the Integral structure *eteology* emerges: *being-in-truth*. This is a direct apprehension of what is true. Instead of reaching a conclusion from rational, analytical processes, the conclusion arrives without cognitive thought, or sequential logic, or "figuring out." It arrives in wholeness, as a single insight, a *gestalt*, in a moment of clarity and knowing that Gebser calls *verition*. The world becomes *diaphanous* (translucent). While mythology stems from *imagination* and philosophy is born from *cognition*, eteology arises from *verition*.

The following poetic tribute to the Integral is taken from the end of *The Ever-Present Origin*:

> the whole,
> integrity,
> transparency (diaphaneity),
> the spiritual (the *diaphainon*),
> the supercession of the ego,
> the realization of timelessness,
> the realization of temporicity,
> the realization of the concept of time,
> the realization of time-freedom (the achronon),
> the disruption of the merely systematic,
> the incursion of dynamics,
> the recognition of energy,

the mastery of movement,
the fourth dimension,
the supercession of patriarchy,
the renunciation of dominance and power,
the acquisition of intensity,
clarity (instead of mere wakefulness),
and the transformation of the creative inceptual basis.

As one explores Gebser's Integral consciousness, it becomes clear that some of his concepts and vocabulary are hard to grasp. A typical passage from *The Ever-Present Origin* illustrates this:

> Only where the world is space-free and time-free, where "waring" gains validity, where the world and we ourselves — the whole — become transparent, and where the diaphanous and what is rendered diaphanous become the verition of the world, does the world become concrete and integral.

The challenge Gebser faced in describing the Integral structure of consciousness is that his readers would be operating primarily in Mental consciousness — most of us expect logical, linear arguments with charts, tables, categories, and stages. But Gebser avoids indulging us in these Mental constructs and writes in a style and language that often seems unfamiliar and even disarming. Jeremy Johnson comments on Gebser's unconventional and masterful use of language:

> … can a work about the integral structure also be an integral expression? This is what makes Gebser's writing, even in translation, so evocative: it achieves a form of literary integrality — the mutual imparting of the integral reality in the reading and the reader, producing an intensified clarity and presence in the text regardless of its perceived mental difficulties.[55]

---

[55]. Johnson, Jeremy. *Seeing Through the World*. Revelore Press (2019).

In the following chapters, we will continue to unpack Gebser's vision as we further explore the evolution of consciousness.

Stepping Back from the Five Structures

Let us now step back from the details of Gebser's work and look from a bigger perspective at what he discovered and what we should take from it. Although scholars and other wise people have been pondering consciousness for a long time, Gebser was the first to identify and describe the fundamental structures he named the Magical, Mythical, Mental, and Integral. These structures have now been widely recognized and adopted by many evolutionary philosophers, and, as we will see in the next chapter, they have now been discovered independently in other fields of science.

Gebser recognized that these structures emerged sequentially as humans evolved from Apes, bringing us right up to today's ecological crisis. In broad brush strokes, this is our story. But perhaps the most important take-away is his vision of the future in which a new intensified structure of consciousness becomes widespread to bring a new era of history for humanity and the planet. Although Gebser is sternly cautionary about the unfolding catastrophe humanity now faces, his message is ultimately one of hope and optimism for the future. I share this view.

Gebser's emerging consciousness is a new way of seeing the world in wholeness, and a new way of being in the world he called the *transparent self*. We will explore these ideas in Chapters 6 and 7, as well as many other aspects of the emerging new consciousness, but first, we will examine two other important theories of evolution — those of Merlin Donald and Arthur Young. We will then bring together the work of Gebser, Donald, and Young into a new and simple way of understanding human evolution — our story — that will help us make sense of the events of today and to chart a course forward into a positive future for life on Earth.

In 3-3 below, we summarize the main ideas in Gebser's theory. See Appendix II for an additional summary.

3-3

Summary of Gebser's Five Structures of Consciousness

Structure and Cultural Stage	Time of Emergence	Dimensionality	Nature of Time	Qualities	Nature of Self	Life Stage
ARCHAIC	Unknown to Gebser	**Zero** Dimensionless	None	**Instinctual** Dreamless sleep	Self and other are one. Non-duality.	Newborn infant. Piaget's* sensorimotor stage
MAGICAL	Unknown to Gebser	**One** Linear	Timelessness Day/night	**Emotional** Ritualistic	"WE" not "I" Clan identity, Separation from nature.	6 weeks to two years old. Piaget's *pre-operational* stage.
MYTHICAL	Late stage is Homeric Greece	**Two** Flat Earth	"Temporicity" Cyclic Calendric Moon cycles, and seasons	**Imaginative** Dreamlike, Irrational Pre-causal, Mythology	Unperspectival Emerging "I" Tribal identity, Death and afterlife	Two to twelve years old. Piaget's *concrete operational* stage
MENTAL	Classical Greece to the present. Literature, philosophy, and mathematics. Culminating with the Enlightenment, classical physics and modern science.	**Three** Spherical Cosmos and 3-D space	"Temporality" Conceptual Linear Arrow of time Clocks	**Logical**, Rational Causal, Dualistic **Philosophy**	Perspectival Egoic, National identity	Twelve years and older. Piaget's *formal operational*
INTEGRAL	First appearance in the early 1900s with Einstein, Picasso, and others. Still emerging.	**Four** Whole Cosmos	Time Freedom: Wholeness of past, present, and future	**Diaphanous** (translucent), Arational. Acausal, Whole **Eteology**	Aperspectival, Transparent, Planetary identity	No Piaget analog. Maslow's self-transcendence?

*Gebser does not mention Piaget

CHAPTER 4

The Macro-stages

Origins of the Modern Mind: Merlin Donald

Although the work of Jean Gebser was largely unknown in North America until very recently, researchers in the fields of cognitive psychology, neuroscience, and anthropology have also been working to understand the evolution of human culture and consciousness. One of the most significant contributions comes from Merlin Donald, Emeritus Professor in the Department of Psychology at Queen's University in Ontario, who is a cognitive neuroscientist with a background in philosophy and anthropology. A prolific writer, he is the author of numerous important papers, as well as two influential books, *Origins of the Modern Mind: Three Stages in the Evolution of Culture and Cognition* (Harvard, 1991) and *A Mind So Rare: The Evolution of Human Consciousness* (Norton, 2001). Early in his career he set out to discover what it was *cognitively* that distinguished the genus *Homo*, and especially humans, from our nearest relatives the apes. While Gebser was not able to provide much detail to the early days of human evolution, Merlin Donald's work fills this void. He has kindly given me permission to use his words extensively in this chapter.

Merlin Donald in 2020

As we consider what it is that separated the first toolmakers 3 million years ago from their predecessors, and from our nearest relatives, chimpanzees, bonobos, and gorillas, the story of Kanzi the bonobo is informative. Kanzi was raised in a research lab at Georgia State University and was trained and studied by several psychologists, principally Sue Savage-Rumbaugh.[56] Kanzi was taught to do many remarkable things, from understanding more than a thousand words, to starting a fire with a lighter and roasting marshmallows (see this on YouTube). One of the most significant claims was that Kanzi could make stone tools. He was taught to bang two rocks together until a flake was produced that was sharp enough to cut a rope and release food. However, according to Merlin Donald, Kanzi "has never improved on his previous performances even after 18 years. Kanzi does not practice his skill or try to refine it; he simply breaks rocks until a piece is sharp enough to cut."[57] Kanzi's stone tools were not even close in sophistication to the oldest stone tools found in Africa that required many steps in sequence to make. In fact, if Kanzi's tools were found by an archaeologist they would not even be recognized as tools.

Kanzi and other apes lack a suite of capabilities that Donald calls *mimesis*, that includes "the ability to rehearse and re-enact past behaviors with the view to improving performance. This is a highly complex capacity, involving major changes in primate metacognitive and supervisory abilities, and memory retrieval."[58] Animals can be trained with rewards like food to do many complex procedural routines, which resemble simple human skills, but they cannot intentionally rehearse and refine skills on their own. Donald proposes that what apes and other animals lack, and what the first toolmakers had acquired, was *voluntary memory retrieval*.

---

56. See the book *Kanzi, The Ape at the Brink of the Human Mind* by Sue Savage-Rumbaugh and Roger Lewin. Wiley (1994).

57. Donald, Merlin. *Evolutionary origins of autobiographical memory: a retrieval hypothesis*. Appears as Chapter 14 in *Understanding Autobiographical Memory*. Edited by Dorthe Berntsen and David C. Rubin. Cambridge University Press (2012).

58. Donald, Merlin. *Mimesis Theory Re-Examined, Twenty Years After the Fact*. Appears as Chapter 7 of *Mind, Brain, and Culture*. Edited by Gary Hatfield and Holly Pittman. University of Pennsylvania Press (2013).

Cognitive psychologists recognize four main memory systems in humans: *procedural, episodic, semantic*, and *autobiographical memory*. Although it is clear that autobiographic memory evolved most recently and is unique to humans, the first three systems have ancient roots and probably evolved together, co-mingled. Animals, such as worms, insects and reptiles, learn patterns of action based on the storage of simple algorithms, or schemas. These are short-term *procedural* memories. Procedures are stored in memory and can be retrieved as templates for action triggered by some reading of the environment.

An example of a procedural memory in humans is the skill of catching a ball. There is a sequence of steps in this procedure — tracking the ball with the eyes, moving the body close to the landing point, and placing the hands around the moving ball. Procedural memories are *general* in nature — they still apply in another similar situation when the details are different. The trajectory of every thrown ball is different, yet the same procedure works for catching any ball. In humans, procedural memory is highly developed, enabling us to do things that require extensive rehearsal, like play the piano or drive a car.

The songs of birds are procedural memories, but birds are also capable of hiding food and retrieving it later. This requires *episodic* memory, which is specific in content. The precise details of how and where the food was hidden must be stored in memory and are later recalled when it's time to find the food. Hiding the food is a specific, detailed *episode* of longer duration than a generalized *procedure*. Though episodic memory is longer-term than procedural memory, episodes are still relatively brief. Most birds and mammals have episodic memory systems, with apes having the highest level of development among animals.

With only episodic and procedural memory systems the world would seem fragmented and meaningless, but *semantic* memory generalizes and binds together procedural and episodic memories into meaningful knowledge. For a human, a typical episodic memory might be that of walking home on a freezing-cold December night in Minnesota, with full details of what you were wearing, whom you were with, what you talked about, and how it felt. Semantic memory would create from this the *knowledge* that winters in Minnesota are very cold. Semantic knowledge is formed from the accumulation of episodic memories, and episodic memories are formed from the interpretation of events by semantic knowledge in an iterative loop. Cognitive researchers are still trying to understand the relationships between all three memory systems, how they interact, which parts of the brain are involved, and how they evolved. However, it is clear that semantic memory became highly developed in the *Homo* lineage, providing the basis for meaningful two-way communication and the sharing of knowledge.

Animals, especially apes, excel at situational (episodic) recall and analysis, but they have difficulty re-presenting an event to reflect on it and assign meaning. Their *semantic* memory system is limited despite having a great deal of specific information. They have difficulty attaching broader *meaning* to representations and inventing symbols to convey meaning. Apes in research labs can learn to recognize spoken words, sign language, and pictures when coaxed with structured rewards from humans; they are able to understand humans remarkably well and make simple responses. But the signs and words they understand are situational and specific, lacking broader meaning. There is little ability to express their own understandings and almost no invention of new symbols and meanings. According to Donald, "their powers of understanding are formidable, but modern apes evidently find it impossibly difficult to construct meaningful expressive actions that encapsulate and reflect their own understanding of events."[59]

This means that apes are limited in their ability to have two-way communication in which knowledge and meaning are exchanged and developed. For this reason, apes have not acquired a knowledge base that grows beyond one lifetime and is passed along to future generations; that is, they have not acquired true culture. They have reached the pinnacle of episodic culture but are not capable of mimetic culture, which includes toolmaking. The ability to generate meaningful knowledge and communicate it to

---

59. From *An Evolutionary Approach to Culture* appearing as Chapter 3 in *The Axial Age and its Consequences*, edited by Robert Bellah and Hans Joas, Cambridge, Massachusetts: The Belknap Press of Harvard University Press, Copyright © 2012 by the President and Fellows of Harvard College.

others is what separates the episodic culture of apes, which does not evolve appreciably over time, from the mimetic culture of the *Homo* lineage that has evolved continuously with an ever-growing knowledge base. According to Donald, what gave our ancestors this ability was *voluntary recall*, which is the ability to *consciously* initiate search-and-retrieval processes from all three memory systems. He explains,

> Voluntary recall is the signature behavioral manifestation of retrieval from memory in humans. Animal memory lacks any capacity for *self-initiated* recall. ... Nonhuman animals can learn skills with appropriate conditioning, but their performances can be retrieved only by external cues that elicit conditioned responses. Voluntary recall, as in self-triggered conscious retrieval, the kind of recall needed to practice a skill, is absent. ... [Humans] possess the neural means of voluntarily initiating a search process, and of consciously controlling that process, to gain access to specific memories, retrieve them, and validate the relative success or failure of the search-and-retrieval process.[60]

When archaic hominins, probably Australopiths, first began making tools about 3 million years ago, they had acquired a new suite of abilities, which Donald calls *mimetic capacity*, or *mimesis*. This was a new stage of cognitive evolution that evolved on top of the episodic cognitive regime that we still see in modern apes. In brief, mimesis entails (1) the ability to rehearse and refine actions and skills, (2) the ability to transmit and maintain skills accurately across generations through imitation, practice, and pedagogy (formal teaching), (3) a group cognitive strategy, or mind-sharing, that produces a growing knowledge base, and (4) a two-way communicative capability. This last point suggests speaking, but the acquisition of complex spoken language was still a few million years away. So, how did the early toolmakers communicate without spoken language?

---

60. *Evolutionary origins of autobiographical memory: a retrieval hypothesis*. Appears as Chapter 14 in *Understanding Autobiographical Memory*. Edited by Dorthe Berntsen and David C. Rubin. Cambridge University Press (2012).

If you could not use words, what could you do to convey meaning to another person and share knowledge? We do this in the game of Charades, a form of mime. This is a wordless way to convey meaning, using only gestures, facial expressions, and body language. And for early toolmakers vocalizations were fair game (unlike Charades), but this was not truly language. Today, we rely heavily on spoken language and written language to communicate, but mimesis is still intact. Body language, facial expressions, gestures, and non-linguistic sounds are all important elements of modern communication.

Mimesis is the conceptual "missing link" between the episodic culture of apes and the language-based oral-mythic cultures of most human societies. With the emergence of mimesis, the transmission and accumulation of knowledge became possible and true culture — *mimetic culture* — began. This also marked the beginning of a brain-culture co-evolution, a positive feedback loop that caused both the brain and culture to evolve together. It was the cognitive revolution in the memory and retrieval mechanisms of the brain — mimesis — that launched human culture; from then on the human brain was immersed in the social milieu of culture, driving its further evolution. This is the basis of the evolution of consciousness, what we have called the culture-consciousness synergy.

Writing in 2012, Merlin Donald elaborates on mimesis:

> Mimesis is an embodied, analog, gestural mode of expression that is inherently reduplicative and collective in nature. It turns the public arena of action into theater. Hence, in a sense, the primal form of distinctly human culture is theatrical, embodied, and performance-oriented. Humans are actors, and initially, in its archaic form, the public face of mimetic culture was a theater of embodied action, manifest especially in the well-documented proliferation of refined skills among archaic hominids.
>
> Mimesis is pure embodied action. It entails a revolution in primate motor skill at the highest level of control. In mimetic per-

formances, action is connected directly to event-perception. Why would such a capacity have evolved? The answer is skill. The primary selection advantage of mimetic capacity is being able to refine and disseminate skills. Early hominids were highly skilled — much more so than apes. This is evident in the archaeological record from finished stone tools that are 2.6 to 3.4 million years old. Conceivably, mimesis might have started to evolve much earlier. Based on stone tools alone, we know that late australopithecines, Homo habilis, and certainly Homo erectus, had the ability to rehearse and evaluate, and thus refine, their own actions. This is direct evidence for mimetic capacity. Rehearsal is essentially a mimetic action: the individual must reenact a previous performance in order to practice and improve it.

Mimetic culture still forms the underpinning of human culture. It persists in numerous cultural variations in expression, body language, and expressive custom (most of which people are unaware of and cannot describe verbally), as well as in elementary craft and tool use, pantomime, dance, athletic skill, and prosodic vocalization, including group display.[61]

Stages of Human Evolution

We have so far explored two regimes of consciousness in higher animals — the *episodic consciousness* characteristic of modern apes and presumably pre-toolmaking hominins and the *mimetic consciousness* associated with toolmaking and culture. In 1991 and again in 2001,[62] Merlin Donald proposed these as the first two stages in the evolution of consciousness. What was next? Recall from chapter one, that three transformational periods stand out in the archaeological record. The first was the appearance of manufactured stone tools about 3 million years ago, and we can now say

---

61. From *An Evolutionary Approach to Culture* appearing as Chapter 3 in *The Axial Age and its Consequences*, edited by Robert Bellah and Hans Joas, Cambridge, Massachusetts: The Belknap Press of Harvard University Press, Copyright © 2012 by the President and Fellows of Harvard College.
62. *Origins of the Modern Mind* (1991) and *A Mind So Rare* (2001).

that this signifies the acquisition of the mimetic consciousness. The next was the Great Leap around 60,000 years ago, evidenced by the appearance of many new kinds of tools, the great diaspora out of Africa that brought humans to every habitable continent on the planet, and the acquisition of complex spoken language. Spoken language was the basis for the oral traditions of storytelling, what Donald calls the *mythic* stage of culture and consciousness.

Cognitively, the development of language probably coincided with the emergence of the fourth major memory system that only humans possess: *autobiographical memory* (AM). Donald states,

> AM is a 'higher' form of memory than semantic, episodic, or procedural memory in that it cuts across these basic categories, integrating semantic and episodic memory material with selected procedural memory content. AM … integrates all this disparate material into a time-based framework built around a virtual self — a truly monumental achievement of cognitive integration.[63]

Autobiographic memory, as the name implies, provides the basis for having a sense of life history — not only the memories of past experiences, but also knowing a continuous life story that is yours only. Once we have this realization, it is inevitable to ask, what happens when I die? And, what am I doing with my life? We can presume that animals without autobiographic memory do not struggle with these very human questions. And, of course, autobiography implies a *self*.

In Chapter Two, I proposed that the self is an important part of the package we call consciousness, and that as consciousness evolves, so does the self. With the acquisition of *voluntary* memory retrieval, launching the mimetic culture, a new stage of self must have emerged. To voluntarily, consciously, intentionally recall a certain memory, or to consciously choose *any* action,

---

63. From *Evolutionary origins of autobiographical memory: a retrieval hypothesis* appearing as Chapter 14 in *Understanding Autobiographical Memory*, edited by Dorthe Berntsen and David C, Rubin. Cambridge University Press (2012).

implies a self *who* is doing the recalling, *who* is choosing. Let us think of the mimetic stage, then, as not only the beginning of culture, and the beginning of our lineage, but also a new stage of selfhood in which action can be *chosen* voluntarily by a conscious entity, whether that be to retrieve the memory of a toolmaking skill so it can be rehearsed and improved or to teach another individual how to find flint for toolmaking. Conscious, reflective choice may be another one of those defining capabilities of the human lineage that emerged 3 million years ago in the mimetic revolution. It would appear that other animals make choices — a chimp may choose a banana over a margarita — but they are reflexive, reactive, unconscious choices driven by short-term needs.

By the same reasoning, it follows that the next broad level in the evolution of the self would correlate with the *mythic* stage of culture. This coincides with the evolution of autobiographical memory and the sense of being a person with a life, who was born and will die. People in this stage of culture — the Upper Paleolithic from about 60,000 years ago until about 10,000 years ago — lived in tribes. The mythical self could exist only within a tribal structure and was not yet fully individuated.

Returning to Merlin Donald, what is his next stage in the evolution of culture and consciousness, after the episodic, the mimetic, and the mythic? He proposes that starting 5,000 to 10,000 years ago humans began to acquire an intellectual capability that he calls the *theoretic* consciousness. This is epitomized by classical Greece and other Axial Age societies, reaching a pinnacle in modern scientific culture. Just as mythic culture rests on spoken language, the theoretic culture is underpinned by written language. Writing created a permanent external storage medium for knowledge, first as clay tablets, then parchment, printed media, and now global digital storage with rapid search and retrieval. This greatly accelerated the growth of knowledge and the development of technology, taking us to our present-day culture.

Let's now summarize Donald's stages of human evolution:

4-1
MERLIN DONALD'S STAGES IN THE EVOLUTION OF CONSCIOUSNESS AND CULTURE[64]

Stage	Species/Period	Novel Forms and Governance
EPISODIC	Primate	Episodic event perceptions, reactive
MIMETIC	Early hominins, peaking in H. Erectus	Skill, gesture, mime, and imitation
MYTHIC	Sapient humans, peaking in H. sapiens sapiens	Language, oral traditions, narrative thought, mythic framework
THEORETIC	Modern culture	External symbolic universe, formalisms, large-scale theoretic artifacts, massive external storage, institutionalized paradigmatic thought and invention

Donald adds the following comment to the summary of stages above: *Each stage continues to occupy its cultural niche today, so that fully modern societies have all four stages simultaneously present.* By now, Merlin Donald's stages of human evolution — the mimetic, the mythic, and the theoretic — should begin to look familiar because they bear a striking resemblance to Jean Gebser's stages and structures of consciousness. They were not contemporaries — Donald was working some forty years after Gebser, at a time when Gebser's work was still largely unknown outside of Europe.[65] He was unaware of Gebser, yet he arrived at very similar stages in the evolution of culture and consciousness based on his knowledge of cognitive psychology, archeology, and primate behavior, fields that scarcely existed in Gebser's time. To arrive at such similar theories from such different starting points is remarkable and supports the validity of this broad description of human evolution. Let's put the two side-by-side:

---

64. From *A Mind So Rare: The Evolution of Human Consciousness*. W.W. Norton and Company (2001).
65. Gebser's *The Ever-Present Origin* was not available in English until 1985.

Gebser	Donald
ARCHAIC	EPISODIC
MAGICAL	MIMETIC
MYTHICAL	**MYTHIC**
MENTAL	**THEORETIC**
INTEGRAL	

Comparing the two, we can conclude:

1. The *mythical* and *mythic* are identical, as are the *mental* and *theoretic*. This is emphasized by the shading above.

2. But we should not take the *archaic* and the *episodic* to be the same thing. As mentioned in Chapter 3, Gebser had virtually no knowledge about our early ancestors, so we may regard his *archaic* stage as more of a placeholder. On the other hand, Donald uses extensive knowledge from research in cognitive psychology, neuroscience, and primatology to elaborate the *episodic* stage and associate it with ape/pre-*Homo* culture and consciousness.

3. We should also not take the *magical* and the *mimetic* to be the same thing. Above, we explored in some detail Donald's case for the emergence of mimesis as the basis of toolmaking and the transmission of culture. I consider this to be a major contribution to our understanding of human evolution. However, I will also argue for the validity of Gebser's *magical* stage as a later development, after the mimetic revolution and before the mythic/linguistic leap, an intermediate stage. Recall that Gebser arrived at the magical structure largely from the extensive studies of primitive peoples in New Guinea and Polynesia by cultural anthropologists of his time. This was a glimpse into the past of cultures isolated from modern civilization that do not exist today. These were largely pre-linguistic and centered on elaborate ritual, shamanism, and nature worship. Although Donald would consider this behavior to be included in the

later part of the mimetic stage, I am proposing that Gebser's magical stage is a valid and distinctly separate stage that built on and followed the mimetic stage. Is there evidence of the magical stage in the archaeological record, as there is for all the other stages? As of now, not in the tool industries, but there *was* a new species of *Homo* that appeared at the right time, roughly one million years ago, with a large brain and advanced behaviors: *Homo heidelbergensis*. It is widely thought that this species gave rise to the Neanderthals, the Denisovans, and early *Homo sapiens*, so we may presume that all of these species represent the magical stage. It was finally the Great Leap about 60,000 years ago that took only *Homo sapiens* a big step further into the mythical consciousness and oral culture.

4. Stating the obvious, Gebser proposed a new stage, the *integral*, following our current stage, while Donald does not. We will return to this point later.

Based on these conclusions, let us now merge the stages of Gebser and Donald by retaining the identical stages and adding the *mimetic* from Donald and the *magical* from Gebser. I will also keep the *integral* because the "next stage" will be a major focus of the rest of the book.

4-2

INTEGRATING GEBSER AND DONALD

Merlin Donald	Jean Gebser	Gebser-Donald Synthesis	Species	Onset (Years Ago)
1. Episodic	1. Archaic	0. Episodic	Early *Australopithecus*	?
			Modern Apes	4.5 million
2. Mimetic		1. Mimetic	Late *Australopithecus*	3.5 million
			Genus *Homo*	2.5 million
	2. Magical		*Homo erectus*	2 million
		2. Magical	*Homo heidelbergensis*	1 million
			Homo neanderthalensis	500,000
			Homo sapiens idaltu	300,000
3. Mythic	3. Mythical	3. Mythical	*Homo sapiens sapiens*	60,000
			Cro-Magnons	30,000
4. Theoretic	4. Mental	4. Mental /Theoretic		10,000
			Axial Age	2,500
			Modern Science	400
	5. Integral	5. Integral		Emerging

We will refer to these major cultural stages and their underlying structures of consciousness as the *Gebser-Donald Macro-stages* of human evolution: the *Mimetic*, the *Magical*, the *Mythical*, and the *Mental/Theoretic*. However, I will propose a further significant modification to the 4th stage later in this chapter. The 5th stage, Gebser's *Integral*, is the subject of Chapters 6 and 7, and from here on I will call this stage the *Planetary* stage of culture and consciousness, for reasons that will become clear. These stages represent the broadest, most macro level of understanding the evolution of human culture and consciousness, hence the name *Macro-stages*. Certainly, more nuance and detail can be added, especially to the more recent of these Macro-stages. In particular, I must acknowledge the extensive work of Ken Wilber and Don Beck, the principals behind today's *integral theory* and *spiral dynamics*, respectively. These well-known frameworks are a more detailed view of human development and cultural memes contained within these Macro-stages.

Before closing our exploration of the work of Merlin Donald, let's return to the question of the 5th Macro-stage, the *Integral* consciousness, the crown jewel of Jean Gebser's theory but without analog in Merlin Donald's work. What does Donald have to say about a future stage for humans beyond his Theoretic stage we now live in? Referring to this as the "fourth transition," Donald writes in 2019:[66]

> I have often been asked whether there might be a fourth transition underway, mediated by the recent explosion of high technology. For over 25 years I have typically taken a pass on such speculation. No more; I am now convinced that a fourth major transition may well be underway, although it is certainly not finished. Rather it is just beginning, and its future direction is uncertain.

But he is not talking about Gebser's Integral consciousness. He is talking about the adaptive changes in the human brain from the ubiquitous digital

---

[66]. From *Self-programming and Self-domestication of the Human Species: Are We Approaching a Fourth Transition?* Appears as Chapter 9 in *Evolution, Cognition, and the History of Religion: A New Synthesis*. Edited by Anders Peterson, et al. Brill (2019).

technologies and the massive quantities of information in which humans are now immersed. Gebser, who died in 1973, ten years before the arrival of the personal computer, could not have known about the digital revolution and the information overload that was soon to come. Referring to this, Donald further states,

> ... it [digital technology] could have very major implications for how we perform our cognitive business. It could also change who we are, and what we aspire to become... If smart technology becomes much more autonomous, and human beings lose what limited control they have enjoyed over their cognitive and social governance, the locus of control will move to the machine domain.

Undoubtedly, the digital culture we live in is shaping our consciousness, as culture has done for 3 million years. We now have massive quantities of information available to us through external storage systems accessible through internet, but is all this information *meaningful* and *helpful*? What benefit does it bring? Ultimately, these technologies, and the sea of information we now swim in, are just more tools in our 3 million-year relationship with tools, and we can either use tools wisely or poorly. I do not want to see humans falling under the control of the machine domain, as Merlin Donald cautions, sold as "improving our lives." I will go with Gebser's view: The next stage of evolution for humanity is a new structure of consciousness co-evolving with a new culture. We will flesh out this vision in the following chapters.

In the next sections, we explore Arthur Young's *theory of process* to gain a deeper understanding of evolution as a universal phenomenon. We will then bring together the Gebser-Donald synthesis from above with Arthur's Young's theory of reflexive evolution to create the *Gebser-Donald-Young Macro-stages* of human evolution. Then, we will have an answer to the question posed at the beginning: What is the phenomenon of humanity, and what is our story?

The Reflexive Universe: Arthur M. Young

Photo courtesy of Arthur Bloch

Arthur Young's theory of process is an all-encompassing theory of the universe and how evolution occurs through the interplay between freedom and constraint within a universal pattern of seven stages. It is a theory of *supra-evolution*. But just a warning, Young's theory of process is at present mostly outside the boundaries of mainstream, widely-accepted science, although Young himself was well grounded in mathematics, physics, and philosophy. His work, like Gebser's, does not fall easily within traditional categories of academic study. But, like all pioneers, Gebser and Young were unafraid to think *way* outside the box as they integrated many fields of knowledge.

Arthur Young was born in 1905, the same year Jean Gebser was born, the year an unknown Albert Einstein published the Special Theory of Relativity. Young attended Princeton to study astronomy, but he soon became more interested in mathematics. During his last two years, he devoted himself to studying the newly developed fields of relativity and quantum physics under the tutelage of Oswald Veblen, one of the foremost American mathematicians at the time. After graduating in 1927, Young wanted to create a comprehensive theory of reality and the universe, one that would include all its processes and history, and also bring together the outer physical world of science with the inner world of human experience, consciousness, and spirituality. This quest would eventually become the *theory of process*, after forty years of work.

After working on this project for two years he wanted to gain more practical experience in the world and searched for an unsolved technological problem to take on. Vertical flight had been studied and attempted unsuccessfully since the time of DaVinci, so in 1929 Young decided to focus his efforts on inventing a helicopter that could fly safely. He began designing, building, and testing small models, and after working alone for twelve years, he finally created a helicopter that didn't crash. In 1941 he reached a manufacturing agreement with the Bell Aircraft Company, and in 1946 the Bell Model 47 helicopter went into production and received the first commercial helicopter license ever granted. The visually iconic Bell 47 (used in the opening scenes of the '70s TV series M*A*S*H) had the longest production run of any helicopter design.

Despite this success, Young was not enthralled with business and was deeply disturbed by the development of the atomic bomb and its use against Japan in 1945, realizing that science lacked a moral dimension to guide choices. He was uncomfortable with the way things were headed in the world and wanted to understand our history and evolution and contribute to a better future for humanity. Young devoted the next fifty years of his life to developing and teaching about the theory of process that explains where humanity is today in our own evolution and how evolution works at every level and time scale of the universe. In 1976 he published the theory of process in his seminal works, *The Reflexive Universe* and *The Geometry of Meaning*. Arthur and his wife, Ruth, founded the Institute for the Study of Consciousness in 1973 in Berkeley, where he worked with his students and hosted symposia until his death in 1995. Anodos Foundation was founded in 1995 to carry on his work. A rich collection of resources can be found at www.arthuryoung.com.

Reflexive Evolution: The Seven Stages of Process

The name *theory of process* can be taken to mean *theory of evolution*, but not just the evolution of life, or the cosmos, or culture — it includes the evolution of every aspect of the outer and inner universe: matter, energy, life, people, organizations, relationships, and consciousness. It is, as such, the ultimate theory of supra-evolution.

Young noticed recurring patterns in the way the physical universe is organized that can be seen in every area of science, and he was able to identify these patterns as evolutionary stages in a completely new way. The theory of process is a new paradigm of the universe that expands the boundaries and methods of science to include non-objective phenomena, such as *purpose*. We now explore some of the key aspects of the theory.

The starting point of process theory is that all process (evolution) occurs in *seven* distinct stages. Furthermore, movement through these seven stages can be thought of as a fall from freedom into constraint, then a turnaround, and a rise from constraint into freedom. Young depicts this movement from freedom to constraint and back to freedom as the *arc of process*, a V-shape like this:

4-3

THE GENERIC ARC OF PROCESS

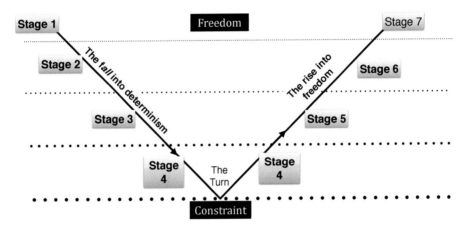

Young calls the first four stages of process the *fall into determinism*. In the arc of process the vertical scale represents *degree of freedom*, so freedom diminishes as evolution proceeds through the first four stages — Stage 4 has minimum freedom and maximum constraint. All evolutionary development proceeds through these four stages, moving from uncertainty (freedom) into certainty (constraint). This fall into determinism is an investment of freedom into *means*, as we'll see from specific examples.

But in Stage 4, evolution changes course with *the Turn*. This is what Young meant by the term *reflexive* — evolution turns back upon itself. The constraint and certainty in Stage 4 provide the means for evolution to regain the freedom that was invested. The evolving entity, referred to as the *monad*, now begins to rise up into more freedom in Stage 5. But this can only happen *if* the lessons of constraint in Stage 4 — the laws of causality and determinism — are learned and applied correctly (this is an important *if* for humanity right now). The monad can then continue regaining freedom as is moves from Stage 5 into Stage 6 and finally into Stage 7, where maximum freedom is regained.

Paradoxically, it is the *determinism* and *constraint* of Stage 4 that make *free will* possible in later stages and provide the means for creativity and agency. This is an important point because it applies to our situation right now on Earth. Philosophers have debated *determinism* versus *free will* for centuries, assuming that they are mutually exclusive; however, Young saw determinism and free will as complementary.

According to Young, this is how all things evolve. He discovered this 7-stage pattern throughout the findings of science — in the electromagnetic spectrum, in the periodic table of the elements, in the hierarchy of molecular bonds, and in the evolution of life.

The Seven Kingdoms of Nature

Let's now apply the generic form of the arc of process shown in 4-1 to a particular domain of evolution, to a particular monad. A good example to start with is the biggest one, at the most macro level: the evolving entity, the monad, will be the entire universe, including life and humanity on Earth. At this largest scale of evolution Young calls the seven stages of evolution the *Kingdoms*,[67] and they are as follows:

1. **LIGHT**
2. **NUCLEAR**
3. **ATOMIC**

---

[67]. These are not to be confused with the taxonomic kingdoms used in biology, Plants, Animals, Fungi, Monera, etc.

4. MOLECULAR

5. PLANTS

6. ANIMALS

7. HUMANS

Let's insert these seven Kingdoms into the arc of process:

4-4

THE ARC OF PROCESS FOR THE UNIVERSE AND LIFE ON EARTH

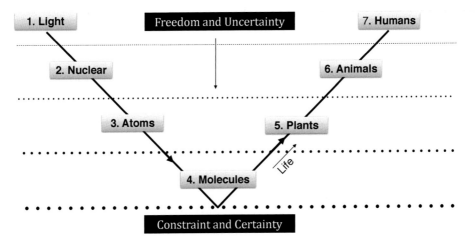

The evolution of the universe begins with *light* as Stage 1 because nothing can "happen" in the universe as we know it without the participation of light. Light can only appear packaged in exact chunks, or wholes, called *photons* — the photon is a *quantum* of light. The photon is sometimes called a "particle" of light, but it is certainly not an actual particle. Photons can seem particle-like because they deliver their energy in a single surge, whose amount is proportional to the frequency of the light. But more importantly in the theory of process, photons are the carriers of *action*. The photon is the *quantum of action*, the smallest possible piece of action. Young realized that *action* is primary in the universe, and not matter or structure. Action is the *means* of process, and all action is *quantized*. This is also true of human actions. Young says, "… action comes in wholes … we cannot have one-

and-a-half or 1.42 actions. We cannot decide to get up, vote, jump out the window, call a friend, speak, or do *anything* one-and-a-half times."[68]

As the carrier of action, the photon is also the initiator of *purpose*. It is the first mover and the source of all action — what Aristotle and later philosophers called *first cause*. To be clear, mainstream science denies both first cause and purpose. There is no scientific explanation for what first cause could be — except God, and that's forbidden in science. In the theory of process, *light* is first cause. Young states that all activity and all interactions of atoms, molecules, and living things are triggered by photons of light.

Mainstream science, taking the *materialist* view, asserts that the universe and everything in it has *no purpose* — there are only "laws of nature." In this view, purpose is something humans invent for various reasons, a cultural construct. Outside of the human drama here on Earth and the things we invent, there is no purpose for anything, including evolution. Life inched forward in tiny steps because of a huge number of random mutations to eventually produce you and me. In this materialist view, the phenomenon of humanity — collectively and individually — has no purpose or significance. But many people find this to be a philosophically slippery slope that leads to the absurdly bleak conclusion that life is meaningless and nothing really matters, except our own pleasure. But Young's universe is alive with purpose. Purpose is essential; it is primary; it is the beginning of all evolution, and it drives evolution forward.

The photon, as the Stage 1 entity, is in a state of maximum freedom, or *uncertainty*. We know from physics that a photon is a piece of light that can't be pinned down in any way: it is uncertain and unconstrained in both space and time — it is impossible to say exactly *where* a photon is at a specific *time*. Furthermore, any number of photons can pass through the same point in space without affecting each other because they are not constrained by space. We *can* specify the instant in time a photon dies, such as when it enters a detecting instrument or your eye.

---

68. All Arthur Young quotes are from *The Reflexive Universe*. Merloyd Lawrence (1976), Anodos Foundation (1999).

The photon is very loosely connected with space and time because it has a velocity. In fact, the *only* thing a photon can do is travel through space at the speed of light, 300,000,000 meters per second (186,000 miles per second). Einstein showed in 1905 that time slows down and nearly stops for an observer traveling at close to the speed of light, and time would stand still for photons moving *at* the speed of light, so time does not exist for photons. The photon is unconstrained in time and space, and it has no mass. This is the highest possible state of freedom, Stage 1 in Young's evolution of the universe.

In Stage 2 of large-scale evolution, as we descend the V in the arc of process, light gives rise to nuclear particles,[69] such as electrons, protons, and neutrons. The birth of nuclear particles has been observed experimentally by nuclear physicists in the process called *pair creation*. Here a high energy photon spontaneously transforms into a pair of charged nuclear particles, such as an electron and a positron or a proton and anti-proton. As the massless photon disappears and nuclear particles appear, the photon's energy becomes mass according to $E = mc^2$. In the early universe massive flashes of light with almost infinite energy created a universe full of nuclear particles, such as quarks and electrons.

Because nuclear particles have mass, they are loosely bound in time — they have lifetimes that end in decay — so they have less freedom than the photon from which they were born. This loss of freedom is invested in *means* — nuclear particles provide the means for atoms to exist. But nuclear particles still have a high degree of freedom because they are not bound in space. They can be described in the wave-based probabilistic language of quantum mechanics as being smeared out over space, but they cannot be precisely located. In Stage 2 there is mass and time but no "hard objects" with knowable locations in space.

Stage 3 is when atomic matter emerges, when these fuzzy nuclear entities join to form atoms. This is a more certain level of existence, with more con-

---

69. It is also not literally correct to call the nuclear entities "particles." In quantum mechanics they are smeared out over space in wave-like fashion, and in quantum field theory they are "squiggles in a field," according to cosmologist Sean Carrol.

straint and less freedom. Evolution is falling further into determinism, as freedom is invested in means. Atoms provide the means to build the physical world we know. The laws of causality, which are the laws of physics, become more constraining for atoms: positive and negative charges must always attract, atoms will always try to be electrically neutral, electrons fill the atomic shells according to the Pauli exclusion principle, and so on. The more complex atoms of Stage 3 are more constrained and less free than the nuclear particles of Stage 2. The 92 types of atoms that exist naturally in our neighborhood (the Solar System) have definite properties and rules of behavior, and these rules determine how atoms can join together as molecules by participating in various kinds of chemical bonds. Atoms provide the *means* for molecular matter to exist.

In Stage 4, molecular matter emerges when the atoms of Stage 3 join together in chemical bonding. Molecular matter in its frozen state — what we call solids — is great for making the stuff of rocky planets and the hard objects governed by the laws of classical physics. This is the physical world we are familiar with. Of course, liquid and gaseous forms of molecules, such as water and oxygen, are also part of the Stage 4 physical world. For us, Stage 4 is the world of material things, such as rocks, rivers, clouds, cars, and even other planets. But Stage 4 does not include *life* — that begins in Stage 5.

Stage 4 of all evolution has the least freedom and the most constraint. The material world around us is highly deterministic, constrained, certain, and predictable: we can get a spacecraft to the Moon exactly as our mathematical models predict. The way things work in the material world is quite locked-in and consistent — rocks always fall downward, and you can be sure that a bus will do serious damage if it hits you. This is cause and effect, or causality. This is the "real world." Through the first four stages of evolution, freedom has been invested to provide means. The highly constrained nature of Stage 4 is also the means for evolution to continue.

Stage 4 is critical because it provides the possibility for *the Turn*, perhaps the most important occurrence in any process of evolution because it allows evolution to continue in a new direction instead of stalling out. This is the point at the very bottom of the V in the arc of process. In the evolution of the universe, this is when inanimate matter gave rise to *life*. The fall into determinism and constraint reverses to become a rise into more freedom and creativity. This *Turn* is a major feature that separates Young's process theory from other evolutionary theories that are linear.

In the large-scale evolution of the universe, represented by the Seven Kingdoms, the Turn occurred here on Earth when life began, and this may have occurred in other places. From the inanimate matter of Stage 4, frozen with constraint, evolution begins to rise up towards more freedom. In Stage 5 a new form of matter emerges — life — activated with a higher degree of freedom. A bacterium (Stage 5) has more freedom than a rock (Stage 4).

Stage 5, as it emerged on Earth, began with prokaryotic life (basically bacteria) that developed the ability to capture the Sun's energy directly through photosynthesis. As life evolved from this simple beginning, an amazing thing occurred about two billion years ago when tiny photosynthetic bacteria were engulfed by larger organisms, a process called *endosymbiosis*. The tiny bacteria became *chloroplasts,* the site of photosynthesis, within the larger organism. This launched the vast lineage of plants as we now know them, culminating with the flowering plants. Plants have the freedom of being able to use the sun's energy directly but are still highly constrained by immobility. But they don't *need* to move because they can capture the sun's energy directly. Young called the 5th Kingdom, "Vegetable," and I have used the name "Plants." Perhaps better still would be "Non-animal life."

The Kingdom of Animals, Stage 6, has the power of *mobility*. Biologists have not arrived at a strict definition of an animal but agree that most animals share these characteristics: they are mobile for at least part of their lives; they are multi-cellular; they are heterotrophic (digesting food internally); they require oxygen and produce carbon dioxide; during embryonic development they originate from a blastula, or hollow ball of cells; and they lack cell walls (as plants and fungi have). The fossil record suggests that the first animals originated between one billion and 650 million years ago.

The recovery of freedom continues in Stage 6 with the mobility of animals, yet animals are still constrained by being restricted to specific habitats. For example, fish can only live in water. Whether a worm, a lizard, or a dolphin, animals have more freedom than plants or rocks, but they are still constrained in their choices and limited in what they can bring about in the world — animals have limited *agency*.

From the Australopiths at the highest level of the Animal Kingdom (Stage 6), the final stage emerges. Young referred to Stage 7 as the Kingdom of *Dominion* or *Man*, but I will refer to it by the scientific name, *Homo,* as we have already established. This new kind of animal began making and using stone tools, which signals the ability to share and accumulate knowledge and skills across generations, transcending genetic inheritance. The acquisition of knowledge and its application, such as toolmaking, brought unprecedented freedom and power to our ancestors, and still does for humans today.

As the *Homo* lineage evolved, it began dominating all other life forms. Young calls the power of Stage 7 *dominion*. In general terms (not Earth-based), the 7th Kingdom refers to a life form that is more powerful than any other on its planet, and so it begins to dominate its biosphere. Today humans "rule the Earth" (or so we may think), but this does not mean we are at the pinnacle of our development — the fact that we are also destroying Earth's ecosystems and driving hundreds of thousands of species to extinction is a big hint that we still have a lot to learn. We can dominate, but we have not yet moved into *dominion*. For Young, *dominion* connoted a certain benevolence, as overseers or caretakers who govern wisely. This is yet to come for humanity.

The seven Kingdoms of nature represent the *ongoing* process of evolution at the largest scale. Pair creation is happening right now, as photons (Stage 1) decay into nuclear particles (Stage 2), which can then form atoms (Stage

3), which can then combine into molecules (Stage 4). Stage 5, when life begins, happened on Earth, but could have happened at other locations in the universe where conditions allow, and if life can continue to evolve, Stages 6 and 7 will unfold. We can suspect, as Young did, that life forms of the 7th Kingdom probably exist at other locations around the galaxy and universe — not just on Earth as humans.

Fractal Evolution: The Sub-stages

Each of the seven Kingdoms of nature also has seven *sub-stages* of evolution, making a repeating pattern within the bigger pattern. This is the *fractal* nature of evolution — it is self-similar across different scales. We can depict this by adding the fractal sub-stages to the main arc of process, shown in Figure 4-5.

4-5

FRACTAL REFLEXIVE EVOLUTION

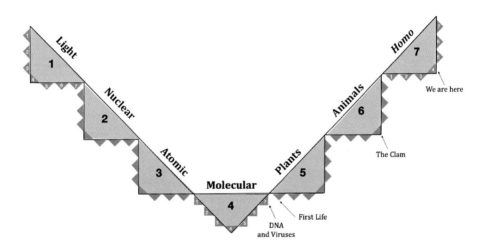

Young sums up this same information — the seven Kingdoms, each with seven Sub-stages — in a seven by seven matrix, shown in Figure 4-6, that he calls *The Grid*.

4-6

ARTHUR YOUNG'S GRID

Courtesy of Anodos Foundation

The Grid: Seven Kingdoms & Their Substages

KINGDOMS ↓	SUBSTAGES →	POTENTIAL	BINDING	IDENTITY	COMBINATION	GROWTH	MOBILITY	DOMINION
1. LIGHT POTENTIAL: No rest mass; outside of space and time; quanta of action; hierarchy	3 degrees of freedom; no symmetry	10^{25} Hz 10^{15} cm 10^{11} eV	10^{22} 10^{11} 10^{7} Cosmic rays	10^{18} 10^{8} 10^{4} Proton rest energy / Nuclear binding energy / Gamma rays	10^{15} 10^{5} 10^{0} X Rays / Atomic spectra	10^{11} 10^{-1} Visible light / Molecular spectra	10^{8} 10^{3} hνkT / Microwaves / Cellular radiation?	10^{4} Hz 10^{6} cm 10^{-10} eV TV and shortwave radio / Animal radiations?	10^{4} Low frequency waves
2. NUCLEAR BINDING: Substance; force; the spell aspect of image, hence illusion; "probability fog"	2 degrees of freedom; bilateral symmetry		Young left these substages unfinished, as work-in-progress					
3. ATOMIC IDENTITY: Acquires its own center; Table of Elements; order creates properties; Pauli Exclusion Principle	1 degree of freedom; radial symmetry Rows of Mendeleef Table	? = electrons per shell	2 One 2 shell HYDROGEN	2 2 Two 2 shells LITHIUM to FLUORINE	2 2 6 One 6 shell SODIUM to CHLORINE	2 2 2 6 6 Two 6 shells POTASSIUM to BROMINE	2 2 2 2 6 6 10 One 10 shell RUBIDIUM to IODINE	2 2 2 2 6 6 6 10 10 Two 10 shells CESIUM to ASTATINE	2 2 2 2 2 6 6 6 6 10 10 10 14 One 14 shell FRANCIUM to ...
4. MOLECULAR COMBINATION: Molar properties; classical physics, determinism	0 degrees of freedom; complete symmetry		METALS Single atom	SALTS Double atoms	METHANE SERIES Non-functional compounds	Functional compounds	POLYMERS chains	PROTEINS Chain with side chains	DNA AND VIRUSES
5. VEGETABLE GROWTH: Self multiplication; the cell or organizing unit; order building by negative entropy	1 degree of freedom; radial symmetry		BACTERIA One cell	ALGAE Many cells	EMBRYOPHYTES Tissue	PSYLOPHYTALES Many tissues	CALAMITES Segmented Larger size	GYMNOSPERMS Mobility of seed	ANGIOSPERMS Flowers
6. ANIMAL MOBILITY: Action and satisfaction; digestion, mobility; choice becomes possible	2 degrees of freedom; bilateral symmetry		PROTOZOA One cell	SPONGES Many cells	COELENTERATES One organ	MOLLUSKS, etc. Many organs	ANNELIDS One chain	ARTHROPODS Chain with side chains	CHORDATA
7. DOMINION CONSCIOUSNESS: Memory of one's own acts leads to conscious knowledge and control.	3 degrees of freedom; no symmetry	?	TRIBAL SOCIETIES (No bodies?) Collective unconscious	——— Self-Consciousness	MODERN MAN ——— Objective thought	Creative genius	CHRIST BUDDHA Mythical kings Mazda?	?

The Grid is a remarkable summation of the evolving structure of nature. Notice that the Sub-stages of the second row, the *Nuclear* Kingdom, are unfinished. It was not possible for Young to fill these in because he did not have sufficient knowledge of particle physics. We know much more today but still lack a robust understanding of the exotic zoo of sub-nuclear entities, such as quarks, neutrinos, mesons, muons, leptons, and the elusive Higgs boson that was finally observed in 2014. Somehow, these must populate the Nuclear Kingdom, but no one has yet pieced this together.

However, the Sub-stages of the 3rd Kingdom (*atoms*) are well understood and reveal themselves in the seven horizontal rows (called *periods*) of the periodic table of the elements, based on the known atomic shell structures (see row 3 of the Grid):

4-7

THE SEVEN SUB-STAGES OF THE 3ʳᴰ KINGDOM (ATOMS)

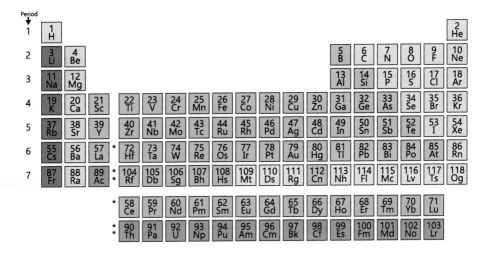

Young discovered the seven Sub-stages of the 4ᵗʰ Kingdom, *Molecules*, by considering the chemical bonds that form between atoms. This is shown in the 4ᵗʰ row of the Grid above and can also be depicted in an arc of process for molecules:

4-8

THE SEVEN SUB-STAGES OF THE 4ᵀᴴ KINGDOM (MOLECULES)

Courtesy Frank E. Barr

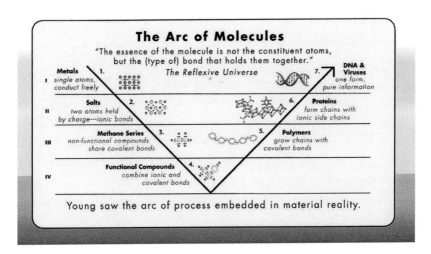

Notice that DNA, the most complex form of molecular matter known, occupies the 7th sub-stage of the molecular kingdom, with the power of dominion (all 7th stages or sub-stages have this power). As such, DNA beneficially manages all the other types of molecules, facilitating protein synthesis and the many other bio-molecular processes of life. Once DNA emerges in the evolution of the universe, life becomes possible. Life, then, begins with the 5th Kingdom and continues to evolve in the 6th and 7th Kingdoms. Young was able to find seven Sub-stages in the 5th Kingdom (plants) and in the 6th Kingdom (animals), again shown in the Grid in the 5th and 6th rows.

For the purposes of this book, the 7th Kingdom is most the important of all the Kingdoms. In the evolution of the universe, this is the point where a life-form begins to acquire the power of dominion. Here on Earth, the 7th Kingdom started with the genus *Homo*, the human ancestral lineage, which we are already quite familiar with.

Like the first six Kingdoms of nature, the 7th Kingdom also has seven sub-stages, and Young was clear that humanity today is firmly entrenched in Sub-stage 4 of our own process of evolution; we are at the *Turn*. Young states that "man is at a state corresponding to the clam in the Animal Kingdom [see Figure 4-5]. Like the clam he is buried in the sand with only a dim consciousness of the worlds beyond." As human consciousness evolves, we will learn to move from *domination* to *dominion*. Instead of being the dominators and exploiters, we will become the protectors of life, and the benevolent caretakers of the planet. This is the sense of what Young meant by *dominion*.

Even though Young was certain that humanity today is in Sub-stage 4, he was not able to clearly identify the first three sub-stages of the 7th kingdom — this can be seen in the bottom row of the Grid that shows little detail. He did not have the knowledge of human evolution that we now have from the field of paleoanthropology. It is now possible to couple this newer knowledge with the work of Gebser and Donald to provide those missing details. We will return to this shortly.

Young viewed evolution as a learning process. In the first four stages, the laws of causality are learned, as freedom diminishes. But then, with the Turn, these laws can be used, if applied correctly, as the means to advance evolution into Stages 5 through 7 where freedom is regained. Process is folding back on itself *reflexively* as the evolving entity learns to use the rules of nature correctly. This is the *reflexive universe*.

Young, the inventor, expresses this beautifully: "There is a fundamental difference between the method of the scientist and the method of an inventor … the scientist, having discovered the law, holds it sacred. The inventor discovers laws, but his goal is to apply them. This involves a change in direction. This attitude of learning how to use the law instead of being confined by it led to my conceiving of law, or determinism, as the agency of free will rather than its antagonist."

Humanity is now in Sub-stage 4 of our evolution, fully immersed in causality and the laws of physics. We have learned these laws well and applied them cleverly to create our powerful technologies. We must now learn to apply our knowledge for *right purposes*.[70] Using the laws of nuclear physics to construct a nuclear weapon is not the right purpose because it threatens the continuation of our evolution. Burning fossil fuels to power our civilization is not the right purpose because it alters Earth's biosphere and climate, endangering many forms of life including humans. The right purposes are life-affirming and healing; they bring value into the world and allow evolution to continue. If we can discern right purpose and act accordingly on a planetary scale, we will move into the 5th Stage of our evolution.

---

70. "Right purpose" is defined more clearly in Chapter 7.

Applications of the Theory of Process

Young shows in his Grid how all things in nature evolve through the seven stages of process, but so do many aspects of everyday life. The evolving entity, the *monad*, can be nearly anything. As a simple example, let's consider the evolving entity to be the relationship between two people who meet and eventually decide to get married. During this process the couple moves from maximum freedom on their first date (they are free to end or continue the relationship) into maximum certainty when the marriage commitment is enacted. The relationship evolves through the first four stages of process, with marriage being Stage 4. But, although things are very locked-in with this stage, the good news from process theory is that the relationship can evolve further into Stages 5, 6 and 7 where freedom is regained. This is the challenge of every marriage (or other committed relationship) — to learn and apply the lessons of causality and determinism in Stage 4; to use the structure and predictability that Stage 4 provides to recover more freedom and creativity. People who are in happy marriages that have lasted for decades have probably accomplished this, and the relationship has evolved beyond the maximum constraint of Stage 4 and into higher stages.

The theory of process has been applied in many human settings. The evolving entity can be a non-profit organization, a business, or a team that works together, but in all cases, the monad moves through the same seven stages that Young discovered. David Sibbet, a leading researcher and practitioner of evolutionary change in the social sector, applies Young's theory of process to help businesses and organizations perform better. Below is one of my favorite applications created by Sibbet and his collaborators.[71]

---

71. The Grove International Consultants: www.thegrove.com/.

4-9

THE EVOLUTION OF SUSTAINABLE ORGANIZATIONS
Courtesy of David Sibbet

Sibbet/Le Saget
SUSTAINABLE ORGANIZATIONS MODEL

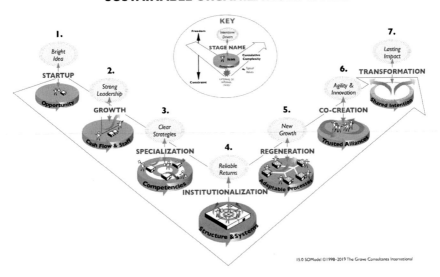

Summary of the Theory of Process (for our purposes)

There is much more to Young's theory of process, but we will close our exploration at this point because we have established sufficient structure upon which to build an integrated stage theory of human evolution. You can explore Young's work further with the resources listed in Appendix 1 and at www.arthuryoung.com. For our purposes moving forward, here is a summary of Young's theory of process:

1. All evolutionary processes and all evolving entities move through *seven stages*.

2. In the first four stages the evolving entity moves from freedom to constraint, then back to freedom in the last three stages. The reflexive turning point in Stage 4 is called the *Turn*. This can be depicted in an *arc of process*, a V-shape with Stage 4 at the bottom (Figures 4-3 and 4-4).

3. When the evolving entity is the *entire universe*, the seven stages of process are called the *Kingdoms*, and these Kingdoms are: light, nuclear, atomic, molecular, plants, animals, and humans (scientifically, *Homo*).

4. Evolution is *fractal*, or self-similar: each of the seven Kingdoms evolves through seven *Sub-stages*, and each of these evolves through seven stages, and so on. The seven Kingdoms and their seven Sub-stages are depicted in the *Fractal Arc of Process* (Fig 4-5), or alternatively in *the Grid* (Fig 4-6).

5. The genus *Homo*, comprising Young's 7th Kingdom, is evolving through seven Sub-stages. We are presently in Sub-stage 4 of our development, and at the Turn, with the *possibility* of continuing our evolution into Sub-stages 5, 6, and 7.

6. Although Young was clear that humanity is now in Sub-stage 4 of our evolution, he was not clear about the details of the first three Sub-stages of our development, as seen in the bottom row of his Grid (Fig 4-6). We can now illuminate that mystery by bringing together the other major pieces we have developed so far in this book, to create a theory of human evolution that we'll call the *Gebser-Donald-Young Macro-stages*.

Before moving on, one note of clarification about Stages and Sub-stages, which can be a confusing matter. Our focus from here forward we will be entirely on the phenomenon of humanity, Young's 7th Kingdom, and the stages of human evolution, as Gebser and Donald described. These would technically be called *Sub-stages* in Young's theory, but for the sake of simplicity, we will now refer to these as the *Macro-stages* of human evolution, or simply *stages*.

Grand Synthesis: *The Gebser-Donald-Young Macro-stages* of Human Evolution

Finally, we are ready to put together the four major components of the human story that we have explored so far:

1. The archaeological record and the knowledge of paleoanthropology.
2. Jean Gebser's stages of culture and consciousness.
3. Merlin Donald's stages of cognitive evolution.
4. Arthur Young's theory of process.

We argued above that the first three of these pieces can be combined to form the *Gebser-Donald Synthesis* with these macro-stages:

1. Mimetic
2. Magical
3. Mythical
4. Mental/Theoretic

Let's now bring in Arthur Young's seven stages of evolution. The Mimetic, the Magical, and the Mythical are the first three stages of human evolution, or, more properly, of the genus *Homo*. Here it would also be tempting to insert the *Mental/Theoretic* structure as the 4th stage of human evolution, but I will now propose an alternate, but inclusive, view based on Young's evolutionary theory.

Young describes the 4th stage of any evolutionary process as being the most constrained and deterministic and having the least freedom. This is an excellent description of life for humans when civilization arose some 5,000 years ago. Not only was the appearance of civilization one of the most prominent events in the entire archaeological record, but it also radically changed the way humans lived, from being nomadic hunter gathers to sedentary city dwellers, from living freely in nature within a tribe to living *indoors* in permanent manufactured structures (buildings) with defined spaces, and locked into large authoritarian social hierarchies.

Five thousand years ago, with the advent of civilization, the world of humans solidified into the *material* world. This is when we began to manipulate material objects on a massive scale, building temples, palaces, and pyramids from 80-ton blocks of stone. Our experience of reality changed from a dynamic, magical natural world to a world of permanence and "stuff," the inert objects in space that we know as objective reality. And, of course, civilization also brought centralized, top-down power structures with ruling elites, minimizing the freedom of nearly everyone, especially the slaves who powered civilization. The radical change in the human condition and lifestyle that coincided with the emergence of civilization 5,000 years ago must be recognized as the onset of the 4th major stage of human culture and consciousness — and our current stage. I will call this the *Material* Macro-stage because of its objective solidness and materiality.

The Material macro-stage is clearly not the same as the *Mental/Theoretic* stage proposed by Gebser and Donald, epitomized by Classical Greece about 2,500 years ago and modern science in the last 400 years. There is quite a discrepancy in time between 5,000 years ago and 2,500 years ago. The Mental/Theoretic structure that emerged in the Axial Age societies can now be recognized as a secondary phenomenon — a sub-stage, if you like — of the Material Macro-stage. The explosion of knowledge and the flourishing of the arts in the Axial Age societies rested on the foundation of Material culture and the supports of civilization. The Material structure of consciousness presents the world to us as objects in space moving predictably through time. This has been our base reality for 5,000 years, superseding the dreamy, pre-causal reality of the Mythical consciousness.

The recognition of the *Material* as the 4th macro-stage of human evolution is consistent with Young's 4th stage of any evolutionary process characterized by maximum constraint, minimum freedom, and full causality. In the evolution of the universe, Stage 4 is the molecular stage when atoms combine to make the solid world of matter — the physical world as we know it, governed by the laws of causality codified in physics. In the evolution of the Animal Kingdom, Stage 4 is the mollusks, like the clam anchored to the floor of the ocean and buried in sand. Analogously, Stage 4 of human evolution — the 4th Macro-stage — is the built world of civilization where both physical and social constraint have reached a maximum. To summarize, we can now recognize the first four stages of human evolution as:

1. The Mimetic
2. The Magical
3. The Mythical
4. The Material

With the addition of the *Material* stage as the 4th macro-stage of human evolution, we can see that these four macro-stages — the Mimetic, the Magical, the Mythic, and the Material — fit appropriately into the unfinished bottom row of Arthur Young's Grid (Figure 4-6); within Young's framework, this also means that we are at or near the Turn, with the *possibility* of three more Macro-stages of evolution:

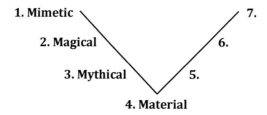

Recall that Gebser considered the European Enlightenment and the birth of science to be the pinnacle of the Mental consciousness, and also the beginning of a "deficiency" or a decline that happens in every stage. He used the term "rational" for this deficient form of the Mental consciousness and referred to this stage of history (the last 400 years) as the "mental rational." This nuance has escaped some interpreters of Gebser, who use the terms "mental" and "rational" interchangeably. They are importantly different in Gebser's view. Ken Wilber also identifies the counterculture phenomenon of the 1960s and '70s as a distinct stage of culture that has come to be known as *postmodernism*.

However, I propose that all of these — the Mental Theoretic, the Mental Rational, and the Postmodern — are *secondary* stages of the much broader and more inclusive *Material Macro-stage* that emerged fully by 5,000 years ago, as shown in Figure 4-10. The Material consciousness is the basis for all that we call civilization, even today; it is also the scaffolding for the Mental/Theoretic cognitive regime that makes science possible. In the next chapter, we delve more deeply into the Material Macro-stage and the modern world, and why its deficiencies now threaten our future on the planet.

We have now married the work of Jean Gebser, Merlin Donald, and Arthur Young and correlated it with the archaeological record, to create the *Gebser-Donald-Young* theory of human evolution, summarized in Figures 4-10 and 4-11. This gives us a simple template for the evolution of humans with four major eras or Macro-stages, with the possibility for three more. This is our story in the broadest brushstrokes. Our focus for the rest of the book will be on Macro-stages four and five, the present and the anticipated, for these are at the heart of our crisis today and our future.

4-10

GRAND SYNTHESIS: THE *GEBSER-DONALD-YOUNG MACRO-STAGES* OF HUMAN EVOLUTION

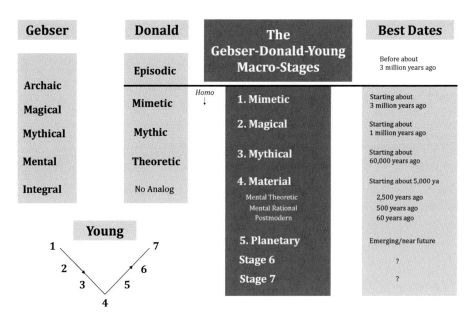

The first four Macro-stages tell our story in the simplest of terms — this is the 3-million-year story of the *Homo* lineage. But what about our future, Macro-stage 5 and beyond that? As mentioned earlier, Macro-stage 5 corresponds to Gebser's emerging new consciousness, the *Integral*. However, I would prefer not to use this term, with apologies to Gebser, because it is used popularly today and has sometimes taken on connotations that are not necessarily true to Gebser. He spent forty years writing and teaching about the Integral consciousness, and occasionally some of that gets lost in contemporary interpretations, so I will instead call Stage 5 the *Planetary* consciousness. The final two chapters of this book will deal with the nature of the Stage 5 culture and consciousness, the *Planetary*, and the emergence of the Stage 5 *Planetary Human*, all very much inspired by Gebser.

What about Stages 6 and 7 in human evolution that come out of Young's evolutionary model? We can only speculate on the nature of these future stages and the longer-term trajectory of human evolution. But I would suggest, as Young did, that the phenomenon of the ascended masters that history records, such as the Buddha and Jesus, fits sensibly as Stage 6 of human evolution. It seems that a rare few humans began attaining this stage thousands of years ago, a phenomenon that has been called *Enlightenment*. The final stage in the story of humanity, then, would be something beyond that, and totally beyond *my* comprehension, so I will not speculate further.

4-11

THE MACRO-STAGES SHOWN IN THE ARC OF PROCESS

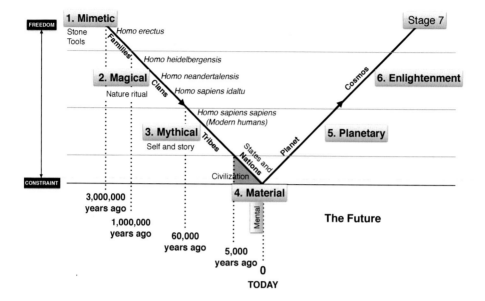

The four Macro-stages are not only the historical eras spanning the last 3 million years; they are also the Macro-*structures* of our own consciousness, and collectively for humanity, manifesting as culture. Human consciousness consists of the Mimetic, the Magical, the Mythical, and the Material structures, each enclosing the previous; and resting on the foundation of the Material structure, is the well-developed Mental/Theoretic structure that governs and dominates our view of the world now. This is the gross anatomy of modern human consciousness and the culture it manifests. In the depiction below created by artist Emily Silver, both the stages of evolution and the structures of consciousness are shown.

4-12

THE MACRO-STAGES AND STRUCTURES OF HUMAN CONSCIOUSNESS

Illustration by Emily Silver

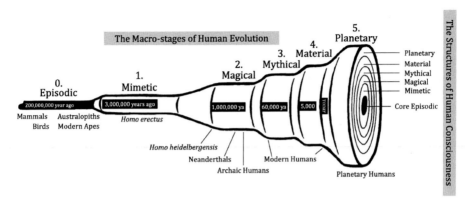

If we humans can successfully navigate this time in our history and evolve into the *Planetary* culture and consciousness, we will have a very promising future. But if we remain mired in the Material worldview, we will slide into dystopia, as we are witnessing today, or perhaps even go extinct, like every other member of the *Homo* lineage. We now consider the future of humanity and how we can achieve the next Great Transformation.

Part II

The Future of Humanity

CHAPTER 5

The End of an Era

We now move to the second major question posed at the beginning of the book: *What is the future of humanity?* But before jumping into that, let's pause, step back, and consider the most important points so far. We've covered a tremendous amount of territory!

- Evolution is a major theme of this book, and a major aspect of the universe. Evolution is much more than what happened to life on Earth. Astrophysicists have detailed the evolution of the universe, and paleoanthropologists have documented the evolution of culture. We can also see evolution in human affairs: in our organizations, businesses, and institutions, in our relationships, and in ourselves. These are not separate things, but one pervasive tendency of the universe we could call *supra-evolution*. We are shifting from seeing the universe as a static, machine-like, collection of *things*, to seeing it as an *evolutionary process* that ceaselessly generates new forms and entities.

- The story of humanity, our story, is the story of the *Homo* lineage that split off from the rest of life some 3 million years ago in Africa. This new life-form had acquired unprecedented cognitive capacities, evidenced in the manufacture of stone tools; this became the basis of true culture: the ability to communicate knowledge and meaning and transmit a knowledge base to future generations. Culture engulfed the large brain of *Homo,* causing it to evolve new conscious capacities, such as language; the evolving conscious brain further advanced culture, creating a positive feedback loop: the *co-evolution of culture and consciousness*. This is what has taken humans so far beyond all other living things. This is the evolution of consciousness.

- The pioneering work of Jean Gebser, Merlin Donald, and Arthur Young, combined with the most current evidence in the archaeological record, shows that the story of humanity unfolded in four major stages or eras that we have called the *Mimetic*, the *Magical*, the *Mythical*, and the *Material* — our present stage. Arthur Young holds that the evolution of any entity, such as humanity, occurs in seven stages, indicating that humanity is far from a finished product — we have three more stages in our evolution. These psycho-historical macro-stages are also macro-*structures* of our own consciousness. The Mimetic, the Magical, the Mythical, and the Material structures of consciousness are all active and important in modern humans, but the Material structure governs at the top level for most humans today.

- The world of chaos and uncertainty we see today reflects the fact that humanity is in transition from the 4th stage of our evolution that is no longer sustainable, the *Material*, into the 5th stage of our evolution in which we will learn to live sustainably and regeneratively on this planet. Jean Gebser, Sri Aurobindo, and Pierre Teilhard de Chardin, three of the greatest evolutionaries of the twentieth century, all foretold of a new consciousness emerging on Earth. Gebser called it the *Integral* consciousness, and we are calling it the *Planetary* consciousness.

As we now begin to envision the future of humanity, we must fully come to terms with the current era we live in, the *Material* era of culture and consciousness. We have been *inside* of it for so long — about 5,000 years — that it's very difficult for us to recognize it and understand it, like the fish that does not see or understand water. Only when a fish crawls ashore and becomes an amphibian can it recognize water; and so too, when a modern human begins to recognize the Material stage of culture and consciousness, then he or she has begun to crawl out of it, onto the shores of Planetary consciousness.

In this chapter we begin by coming to grips with the Material consciousness and culture that we have been swimming in for 5,000 years, and then we will explore the shift into the fifth stage of human evolution. By understanding and clarifying the nature of the Material regime, and by fully owning it, we can begin to get outside of it, to put it behind us, and evolve naturally into Planetary culture and consciousness.

Owning the Material Stage

Let us now come to terms with the Material macro-stage of culture and consciousness we now live in. But brace yourself because some of this is not pretty. We are talking about the era of the *built world*, or the "man-made world" in older parlance, the Age of Civilization. Riane Eisler identified this era more than thirty years ago in her book, *The Chalice and the Blade*, as the *dominator culture*, contrasting it with *partnership culture*.[72] David Korten, mincing no words, calls this the *Imperial* era:

> Imperial Civilization organizes around hierarchies of domination and exploitation. The few at the top use their power to control and exploit the gifts of nature and the labor of the many in a ruthless competition for ever more power. The ultimately self-destructive competition attracts those among us who are driven by immature and narcissistic egos that feed on crushing enemies, displays of opulent consumption, and grand monuments that bear their names … For all the seeming recent human advances in democracy, civil rights, and gender equality, Imperial Civilization prevails to this day under the global rule of transnational financial markets and corporations.[73]

As we've seen, the onset of the Material stage is dramatic in the archaeological record. Seemingly out of nowhere, megalithic structures began to appear, first at Göbekli Tepe starting around 11,500 years ago, and later at many other sites around the world. For millions of years our ancestors

---

72. *The Chalice and the Blade*. HarperCollins (1987). *The Partnership Way*. HarperCollins (1988).
73. From *Imperial Civilization*. https://davidkorten.org/home/great-turning/empire/.

were nomadic hunter-gatherers who left little trace of their presence, except for stone tools and a few of their bones. Then, in a relatively short time, massive stone structures began to appear at many locations around the world. This is a landmark development in the story of human evolution that signals the transformation into a major new regime of culture and consciousness.

The megalithic structures of Göbekli Tepe may be taken as the first signs of the built world, but soon after, proto-cities like Jericho and Çatalhüyük began to appear in the Middle East, where large numbers of people lived in permanent dwellings. People were no longer living in nature but in artificial environments, in spaces defined by walls, ceilings, and floors. There was now a clear distinction between indoors and outdoors, fully separating humans from nature, and humans from other humans. The paradigm of separation began to emerge at this time: artificial structures partitioned space, and spaces were owned, creating the separation of *mine* versus *yours*, *me* versus *you*, and *us* versus *them*.

The relatively sudden appearance of the built world in the archaeological record reflects much more than just a new ability to move big, heavy objects and stack them up. It represents a new psychological regime, a new phenomenal reality. Material consciousness generates a reality of matter, of things, separated in empty space, a world of objects and forces that humans could increasingly exploit. The concept of empty space probably did not exist for our *Mythical* and *Magical* ancestors who saw a dream-like world where space was a plenum filled with a menagerie of energetic and physical entities. In the *Material* regime, power, force, and coercion become major themes, and these apply to both nature and to other people. The power of humans over things — both non-living and living — reached a new height.

The first megalithic structures and proto-cities were small in scale compared to the great civilizations that emerged a few thousand years later; and the early proto-cities were distinctly missing something that the true cities of civilization had: *centers of power*. In the excavated layers of early Jericho and other proto-cities, there are no signs of social hierarchy or a ruling class. There were no palaces or royal quarters but instead, very homogeneous dwellings. Undoubtedly, the construction of megalithic structures and proto-cities required both organization and manpower, but these early societies seemed to have been largely egalitarian. Apparently, everyone pitched in to do the work, and decision-making was not centralized or authoritarian. Furthermore, skeletal remains show that nutritional levels were very similar across whole populations, meaning that no one was eating much better than anyone else. These were structurally flat societies with very little hierarchy, and the abundance of female figurines suggests that they were also matriarchal.

But this changed dramatically between about 6,000 and 5,000 years ago when the first true civilizations appeared in Persia and then Egypt, and soon after in what is now India, China, Crete, Peru, and Mesoamerica. In all of these early civilizations, there is abundant evidence of centralized power and authoritarian rule. The magnificent city-states included monumental architecture, such as palaces, temples, and pyramids with lavish quarters for royalty and large chaotic districts where the commoners lived in poverty. Emperors, high priests, god-kings, and their cadre of elites were at the top of the social hierarchy, and they owned all the power and resources. At the bottom were the slaves who built the great monuments and powered these societies. Wealth and poverty had never before existed in the story of humans, but with the advent of civilization, wealth and power became concentrated in the ruling elite, while the masses toiled in poverty.

The centralized power structures that emerged with civilization were possible because of something else that had never before existed: organized armies of soldiers with powerful weapons who could enforce the wishes of the rulers. Nothing like a little torture and some public executions to improve morale!

For the first time on Earth, there were fortified walls around cities, along with sophisticated weaponry and armor. Organized warfare was invented with civilization. When you put powerful rulers with an appetite for more territory together with armies at their command, you get *empire*. The history of the last 5,000 years has been an endless succession of powerful rulers and dynasties seeking more territory, from the great empires of Babylon, Persia, and Rome to Genghis Khan, Napoleon, and Hitler. The concepts of territory, boundaries, and empire also imply ownership — land, objects, and people as property — something that never occurred to our nomadic ancestors nor to most indigenous people.

We can begin to see the primary characteristics of the Material macro-stage of culture and consciousness. First, is the *physicality* of the built world, a *world of objects* that could be manipulated. Not only were people building impressive permanent structures, they were altering the natural environment through deforestation, agriculture, and urbanization. Secondly, the ever-growing *separation from nature,* and a loss of sacred regard for nature, heightened by living in an artificial world of buildings and cities. And thirdly, *centralized power structures* with ruling elites at the top who held most of the wealth and power, leaving most people in a state of scarcity and servitude. These are not three separate things; they are interrelated manifestations of the same consciousness, and all three of these are alive and well today. We are still living in the Material regime of reality.

Civilization was not all bad. It brought with it many new innovations, such as metallurgy and the use of bronze and steel; a new generation of metal tools brought an end to the stone age. Systems of writing emerged in most civilizations (the Incas are one exception). This was the beginning of permanent external storage of information, starting with clay tablets and progressing to computer memory. Written language catalyzed the explosion of knowledge and creativity that was the Axial Age about 2,500 years ago. Language, both spoken and written, remains one of the most unique and important of human capacities. Just as spoken language gave birth to the Mythical culture, written language underpinned the Material culture and consciousness. Writing is, literally, a *material* form of information with relative permanence. Large-scale agriculture was another major innovation that made it possible to feed large populations, but the grain-dominated diet actually produced a decline in nutritional levels for the common people, while the royalty enjoyed fresh meats, vegetables, and fruits.

There is widespread agreement that classical Greece and other Axial Age societies flourishing around 2,500 years represent the birth of the "modern mind," Gebser's *Mental* consciousness, and Donald's *Theoretic*. However, I proposed in the last chapter that the Mental/Theoretic consciousness was a *secondary* development resting on the scaffolding of civilization and the *Material* consciousness that emerged thousands of years earlier. I have argued that the Material stage is the major stage that stands out in the archaeological record and signals a sea change in how humans lived and viewed the world, while the Axial Age did not bring anything so radically new. Rather it amplified things that were already in play — it was a fruition. Writing, for example, had been around for thousands years by the time of the Axial Age when it ripened into the tragedies of Euripides, the works Lao Tzu, and the Upanishads. These were maturations more than innovations. In contrast, the emergence of the Material culture and the first civilizations thousands of years earlier brought an entirely new way of life for people, after millions of years living freely in nature.

Two thousand years after the Axial Age — about 400 years ago — there was a second irruption of the Mental consciousness, coinciding with the scientific revolution and so-called modernity. Gebser referred to this as the *Mental rational* consciousness, or simply the *rational*, and considered it to be a "deficient" form of the Mental consciousness. By the time of the Industrial Revolution there were increasing signs of decay and deterioration, such as large-scale pollution, child labor in dreadful factories, and the genocide of indigenous peoples throughout the New World by European colonizers. Gebser saw the events of the twentieth century, with two devastating World Wars and the atomic bomb, as signs of the end of the Mental stage, signaling the emergence of the next structure of consciousness, the *Integral*.

I am saying that the *Material* macro-stage of culture and consciousness that began 5,000 year ago has important *sub-stages*, such as the *Mental* consciousness that first emerged 2,500 years ago and the *Mental rational* that appeared 400 years ago. Let us add to that another important cultural occurrence, the so-called *counterculture* of the 1960s and early '70s, what

Ken Wilber and others call the *postmodern*. To be clear, let's summarize the Macro-stages and their sub-stages as follows:

Macro-stage of Evolution	Approximate Date of Onset
1. ***Mimetic***	3 million years ago
2. ***Magical***	1 million years ago
3. ***Mythic***	60,000 years ago
4. ***Material***	5,000 years ago (First Civilizations)
(a) Mental (Gebser)	2,500 years ago (The Axial Age)
(b) Mental Rational (Gebser)	400 years ago (Scientific Revolution)
(c) Postmodern (Wilber)	50 years ago (The Counterculture)
5. ***Planetary***	Emerging (Gebser's ***Integral***)

As we consider the sub-stages of the Material, shown above, I will suggest that we can actually identify *seven* such sub-stages within the Material, in accord with Arthur Young's fractal evolution — where every stage has seven sub-stages. For our purposes here, I will not flesh this out further, except to suggest that in 2020, as I write this, we are in a new sub-stage of the Material, after the postmodern, that we might call the *Post-truth/virtual*. This is the result of the awesome and frightening capability of digital technologies to create "fake" or virtual realities in ways that are indistinguishable from actual reality. It is now quite possible to manufacture reality for people, or to put it another way, to manufacture truth. Hence, post-truth. Today, especially in the U.S., there is a sense of alternate realities existing side-by-side, some based on intentional distortions, disinformation, and lies. As we wake up to recognize this, we will recover the importance of seeking and apprehending truth, of discerning what is real. The most relevant question at this time in history may well be, what is real?

One way or another, the crisis of our time signals the end days of the *Material* culture. Georg Feuerstein, writing in the late 1980s, articulates Gebser's concerns about the decline of civilization:

> As he [Gebser] sees it, ever since the Renaissance we have been on a single downhill rail, taking us to ever more extreme manifestations of what he calls the *rational* consciousness. Our civilization, enamored of the apparent success of our science and technology, has for the past three hundred years celebrated *ratio*, or reason. Gebser, airing the feelings of some of the best minds of our time, debunks this naïve view. For him, *ratio* is the straight-jacket into which the mind is forced when it fails to be motivated or awed by anything larger than itself. Thus compressed into itself, ratio is at the same time hostile to itself and the world: it is the very same mind, or style or relating to itself and the world, that is today ransacking our planet's natural resources, polluting air, land, and water, proliferating nuclear arms, killing off entire animal species, and broadcasting images of strife and violence as models for the younger generation ... Rather than discarding our civilization's technological accomplishments, we must find a new equilibrium on a higher level. That equilibrium would be based on the realization of an intensified and not merely "expanded" consciousness.[74]

Today we are still very much steeped in *rational* consciousness within the *Material* worldview. When empirical science emerged in the 1600s, it elevated thinking, reasoning, and logic — rationality — to the highest aspirational levels, and this remains true today. We still believe that any problem can be solved by the human intellect, and indeed, the dazzling technologies that continue to emerge seem to be proof of this. It was the combination of the *Mental rational* consciousness resting on the foundation of the *Material* that gave rise to classical physics, arguably the crowning achievement of modernity. Newton's mechanical universe was the world of material objects in space described perfectly by the clean intellectual symbolism of

---

74. Feuerstein, Georg. *What Color is Your Consciousness?* Robert Briggs Associates (1989).

mathematics. But as we will see in the next section, this perfectly logical, mechanical view of the world collapsed for the physicists around 1900.

The thinking, analyzing mind partitions the whole world, including its people, into categories with labels. The Mental consciousness is head-centered: we are admonished to "use your head," and we hope to "have a good head on our shoulders." We think things through; we figure things out, and we seek explanations (also called theories). Most modern people think they reside in their heads, in the mental realm. On the other hand, most indigenous people and others who are not governed by the Mental consciousness feel they reside in their heart and gut.

Even though we are firmly entrenched in the Mental regime of culture and consciousness, as Gebser and Donald have demonstrated, we are more broadly in the *Material* macro-stage. All of the original primary characteristics of the Material paradigm are still with us today: the built-world; our disconnection from nature (now causing massive environmental destruction); hierarchical societies that concentrate power and wealth; the world as a collection of objects in space; separation and scarcity; and the use of force to bring about results (for example, war).

For most of us who learned "World History" in school, the story of humanity begins with the first written records and the advent of civilization some 5,000 years ago in Mesopotamia and Egypt. We have traditionally regarded humans before this as "pre-historic," thinking of them as "cave men" barely above the level of animals. But in recent decades, research in paleoanthropology and cognitive psychology has produced a much deeper understanding of human history and evolution, going back *three million* years, and we can now see that what we previously thought of as "history" is simply the most recent era of history, the 4th stage of human evolution, the Material stage of culture in the bigger story of humanity.

The traditional version of history taught in school is more accurately *political* history. It is an accounting of the continual succession of authoritarian regimes with absolute power that have dominated human affairs for 5,000 years. The names and faces change, but the basic script is the same.

The tools of war have changed from swords, spears, and arrows to nuclear bombs, cruise missiles, and disinformation, but these are mere details. The plot is the same, so it's no wonder many people think that "history repeats itself." This is all too true if your knowledge of history goes back only 5,000 years, but obviously, we are taking a much bigger perspective in this book. From that bigger perspective we see that humans have been evolving all along, and we are still evolving; we are not stuck in an endlessly repeating cycle of domination, war, and destruction. *It's only a stage.*

Perhaps the most prominent feature of the Material stage, and any 4th stage of evolution, is the lack of freedom and the high level of constraint. This is pervasive throughout the history of civilization. For the average person 4,000 years ago, options were limited and consequences were severe for wrong behavior. Today, the constraints are not so much from authoritarian rulers and governments (though these still exist in many places), but from new and more subtle forms of constraint that have emerged because of our electronic technologies. The use of radio, television, and now dazzling new screen technologies has opened up many new ways to control people, often without them even knowing it. Today's powerful screen-and-sound technologies are direct portals into the brain that can strongly influence how people see the world, what they aspire to, what they want to look like and be like, and what they want to buy. And the rampant consumerism pedaled so relentlessly today often results in heavy burdens of debt, a brutal form of constraint, and a tragic loss of freedom. Today the *average* American household debt has reached $205,633, which includes $5,331 in credit card debt.[75] How free is that?

Our political processes have been hijacked by manipulators of digital media to create fraudulent realities for large numbers of people based on misinformation and lies. Though most people would claim they are free to make their own choices, we sadly underestimate the power of electronic screen technologies, social media, and the science of advertising to control people. Otherwise, why would advertisers and political groups spend billions of dollars trying to reach people in this way? We will never be able

---

75. https://shiftprocessing.com/credit-card-debt/.

to acquire more freedom until we start to recognize the constraining, controlling influences we are immersed in because we are so enamored with the latest screen technologies. No one forces this on us. It's a choice that can become an addiction.

After 5,000 years of Material culture exploiting, dominating, and destroying, we have maxed out our planet's life support systems. We have poisoned the waters, the air, and the food chains; our exploding population is destroying natural habitats at a staggering rate and driving thousands of species to extinction; and we are altering our climate on a planetary scale. People in the U.S. today are unhealthier than ever before by almost any measure. Obesity is widespread, suicide rates are climbing, opioid addiction is rampant, mass shootings are routine, and average life expectancy is down for the first time ever.

Something seems to be very wrong. Most people have a sense of this, but they have no idea what's going on or how to make things better because they are merely trying to survive. This is not a pretty picture, but it's exactly what the Material consciousness has produced for us after 5,000 years. While it may have served us well at times in the past, it no longer can.

The vision of an evolving new structure of consciousness is hopeful, but we must not think it is guaranteed. If we cannot recognize the Material stage and its now-destructive tendencies, then things do indeed look bleak for humanity. The power structures that exist in the world today — economic, political, and cultural — seem intractable, unchangeable, and incapable of solving our mounting problems. More technology will not be the answer, and we will not think our way out of the material world we created. The only hope is for a new consciousness to emerge on a large scale, so that we will see the world and our place in it in a new way, as a planetary whole. Only when we recognize and own the Material regime will be able to move into something new. This may sound impossibly difficult, but it's exactly what physicists had to do a hundred years ago.

The Story of Physics

The ascendance of the Mental consciousness over the last 400 years and the crisis we are now experiencing, closely parallels the story of physics and how it was transformed. I will suggest that what happened in physics about a hundred years ago — a crisis, a collapse, and the birth of something radically new — was a preview of the massive transformation we are now in the midst of. But the story of physics has a good ending. What emerged from the crisis was a radically new worldview, the quantum worldview, that is completely analogous to the emerging new consciousness today.

Classical physics, originating with Galileo and Newton in the 1600s, is the quintessential manifestation of the Material/Mental structure of consciousness, giving rise to the worldview known as the mechanical universe. Whether we are physicists or not, we all still see the world as predictable, sensible, and machine-like. There is no magic.

At the peak of the Mental consciousness in eighteenth and nineteenth century Europe, classical physics and its mechanistic paradigm were spectacularly successful at describing the known world and producing the technologies that powered industry and commerce. It was considered by many to be infallible, if not perfect. There were some who questioned this and pushed back. The poets and writers of the Romance period — Blake, Thoreau, Whitman, and Emerson, for example — railed against the Newtonian universe and the Industrial Revolution, seeking a reconnection with nature; but they were no match for the power and profits of the mechanistic paradigm and the Material culture.

Classical physics views the world as a collection of "solid, massy, hard, impenetrable, moving particles" — Newton's words — that respond predictably to forces. Newton expressed this mathematically in his famous second law of motion, $F = ma$ (total force is equal to mass times acceleration). The mechanical universe is a deterministic world, like the game of billiards, where every event is caused by something before it, which was caused by something before that, putting free will and self-responsibility into question. The canon of mathematical equations that embodies classical physics is in fact so effective that it enabled us to put a man on the moon. It works perfectly for the "macro-world" we are familiar with, and we use it all the time, for every skyscraper that stands and every jetliner that takes off and lands safely, among many other things.

By the early 1880s it was widely believed that classical physics had successfully explained *everything that could be known* about the physical world, and there was nothing left for science to do. Many young people were advised against a career in science because it was thought to be a dying field. The inventions of Edison, Tesla, and Bell were considered to be so impressive that many people believed that *no more new technology was possible* — that everything that could possibly be invented had been invented! Today, in hindsight, we can see how short-sighted, unimaginative, and just flat wrong this was. This hubris, this arrogance, illustrates the growing deficiency in the Mental structure that was becoming evident by the end of the nineteenth century, according to Gebser. But a severe correction was about to take place in physics.

In 1886, just when classical physics was at its peak and Western civilization was riding a wave of confidence and self-satisfaction, a simple discovery brought down the house of physics cards. A German physicist named Heinrich Hertz observed that light, when shined on certain metals, could produce electricity. He had discovered the *photoelectric effect*, the very thing that makes today's solar-electric panels work. However, further experiments produced results that could not be explained by any known physics. The mechanistic paradigm of classical physics could not explain how atoms could turn light into electricity, and in fact, it could not explain the behavior of atoms at all.

The inability to explain this discovery became a crisis in physics; instead of thinking they knew *everything*, now physicists wondered if they knew *anything*, and for the next 20 years the photoelectric effect was considered the greatest unsolved problem in science. Finally, in 1905, a young unknown patent clerk named Albert Einstein solved the problem by building on the recent work of Max Planck and invoking something that the Mental consciousness cannot grasp — the *wave-particle duality*. For centuries scientists had debated whether light was a wave *or* a particle, but Einstein proposed that light had to be a *particle* with a *wavelength*, that is, both a wave *and* a particle. This is the photon. The Mental consciousness, which separates and categorizes, logically demands that light must be *either* a wave *or* a particle. But a photon is *both* a wave *and* a particle.

Most of us today, more than 100 years later, probably can't really picture how something can be both a wave and a particle. But if we do the physics, we'll see that particles of light — photons — must exist; today's solar energy industry depends on it. Yet experiments in high school physics classes[76] show that light must be a wave. By bringing the photon into the picture with its wave/particle duality, Einstein discovered a simple explanation for the photoelectric effect, and this opened the door for the sweeping transformation that physics was about to undergo. It was this explanation of the photoelectric effect that garnered Einstein a Nobel Prize in 1922.

But this was only the beginning for Einstein and only the first sign that the Mental consciousness was being pushed beyond its limits. Also in 1905, the 26-year old published his landmark paper on the *Special Theory of Relativity* that shattered all sensible notions of time, space, and mass and introduced $E = mc^2$. Nearly everything sacred in physics was overturned by the rubber-like reality of special relativity, and the repercussions spread throughout Western culture. Picasso's use of Cubism around 1910 and Stravinsky's radical *Rite of Spring*[77] both reflect the rupture in the fabric of the Mental consciousness that relativity had torn open.

As Einstein became a celebrity, he spent the next ten years creating his *General Theory of Relativity* that treated gravity as curved space and time. Both of his theories of relativity — special and general — describe a very different world from the one we know. The warping of space, time, and mass is beyond the sensibilities of our Mental consciousness and its limited conception of what the "real world" is. Theoretical physics was revealing a reality that defied common sense and reason, yet many experiments over the last 100 years have validated this reality.

---

76. Such as double-slit interference experiments.
77. Reportedly causing riots in Paris when it debuted in 1913.

Einstein's theories of relativity were merely the opening act in the revolution that transformed physics during the first decades of the twentieth century. Working independently from Einstein, a group of brilliant young mathematical physicists tackled the problem of explaining the behavior of atoms, something else classical physics was unable to do. Through the collective efforts of Niels Bohr, Werner Heisenberg, Erwin Schrödinger, Paul Dirac, Wolfgang Pauli, and many others, a completely new physics emerged by about 1927 that successfully described reality at the atomic level. This has come to be called *quantum physics*, and we now know that it provides a more accurate and complete description of reality than the mechanistic view of classical physics. However, the implications of quantum theory were disturbing, especially to Einstein.

Building on the wave-particle duality pioneered by Planck and Einstein in the early 1900s, quantum theory treats the particle-like electron in an atom as a *wave* described by a mathematical equation, the *Schrödinger wave equation*, that gives probabilistic results. Quantum theory revealed a world that was no longer solid and certain, but instead fuzzy and probabilistic, with limits on certainty that Heisenberg expressed in his *uncertainty principle*. Quantum theory destroyed the mechanistic paradigm of hard objects in empty space that interact through causal forces. The Newtonian machine had broken down, and a new paradigm of reality had emerged, one that we are still trying to understand today.

David Bohm, one of the preeminent quantum physicists of the twentieth century, sums up the main features of quantum theory in the following four points, all of which radically contradict the mechanistic view of classical physics:[78]

> 1. *Discontinuity and quantization.* The states of systems are *discrete* rather than *continuous*. For example, when an electron jumps from one atomic energy level to another it does so without passing through an intermediate state (the "quantum leap"). If your car behaved like this, your speed would jump from, say, 11 mph to 19

---

[78]. I am paraphrasing from *Wholeness and the Implicate Order.* Routledge (1980).

mph without passing through any speeds in between. In this imaginary quantum car, speeds (or energy states) are *quantized*. This would make for a very jerky ride, but, luckily, things the size of real cars have *continuous* energy states.

2. *Wave-particle duality*. Entities, such as electrons, can appear as being wave-like, particle-like, or something in between depending on the environmental context within which they exist.

3. *Statistically revealed potentialities*. Every physical situation is characterized by a wave function (a vector in Hilbert space for you math majors) that is not directly related to the *actual* properties of an object, event, or process. It is a description of the *potentialities* within the physical situation. The wave function gives only a probability for the actualization of different potentialities but cannot predict what will happen in detail for any individual observation.

4. *Non-causal correlations and the paradox of Einstein, Podolsky, and Rosen (EPR)*. Events that are completely separated in space and have no possibility of causal interaction are still connected non-causally. This is also known as *non-locality*. For example, if two electrons that have been living together in some atom are separated to a very great distance, they somehow remain in touch with each other. If one of them is disturbed, the other one shows an immediate effect — but not from any *cause* in the classical sense. This non-causal, non-local connection was so disconcerting to Einstein that he called it "spooky action at a distance" and invented the thought experiment known as the EPR paradox to show that this was impossible and incorrect. However, it has now been shown repeatedly that this phenomenon, now popularly called *entanglement*, is both possible and correct.

Relativity shatters the classical notions of time, space, and mass, but it still requires continuity, causality, and locality. Quantum theory goes much further in redefining reality because it requires non-continuity, non-causality, and non-locality. This is an entirely different order of reality, where things are invisibly connected. Relativity and quantum theory have never been fully reconciled, but they have found a peaceful co-existence because one works in the macroscopic world and the other works in the microscopic world.

Classical physics was a product — arguably the greatest product — of the Mental Theoretic consciousness. It put a man on the moon. But it is based on the mechanistic order in which, according to Bohm, "The world is regarded as constituted of entities which are *outside of each other* in the sense that they exist independently in different regions of space (and time) and interact through forces..."[79] Bohm calls this world that we take in with our senses and our measuring instruments the *explicate order*. This is the world of causality and locality we know so well, the world of separate objects, the reality of the Material consciousness. We also see *ourselves* as separate from everyone and everything, which is the root of the loneliness and alienation that existentialist philosophers emphasize. One of the primary operating assumptions of the explicate order, classical physics, and the Material structure of consciousness is *separation*, or what Bohm calls *fragmentation* (more on this in Chapter Six, *Wholeness*).

From the mechanistic perspective of classical physics and the Mental consciousness, the human body is a machine made of many separate parts. Likewise, all of nature — the whole universe — is thought to be a machine: the *mechanical universe*. A machine, such as a car, can be disassembled into parts and then put back together to make the same machine. A machine can be *reduced* to its parts. When we apply this view to the whole world, it's called *reductionism,* and this is a core assumption of the Mental consciousness and much of mainstream science.

But living things are not machines — they cannot be taken apart and then reassembled. Within our body, every unit, whether a molecule, a cell, or an organ, is connected in some way to every other, and the whole thing behaves more like Jell-o than a collection of parts. If you poke a mound of Jell-o in one place, the whole thing wiggles. The human body, or any life form, is a fully integrated *whole* system that wiggles.

---

[79]. *Wholeness and the Implicate Order.* Routledge (1980).

Living things are *wholes* that are embedded in greater *wholes* that we call ecosystems, and ecosystems are embedded in the biosphere, which is tightly connected with the Sun. None of these exist independently. Bohm states, "Ultimately, the entire universe (with all its 'particles,' including those constituting human beings, their laboratories, observing instruments, etc.) has to be understood as a single undivided whole, in which analysis into separately and independently existent parts has no fundamental status."[80]

This is an entirely new understanding of reality that Bohm calls the *implicate order*. Our familiar world of solid, separate objects in three-dimensional, rectangular space, governed by logic and causality, is a highly filtered version of the underlying *implicate order* that quantum theory points to. From the perspective of quantum theory we are now beginning to see nature, and the whole universe, as an organism, rather than a machine. Goodbye mechanical universe, hello organismic universe.

In just 25 years, from 1905 until about 1930, the field of physics was completely transformed. Relativity and quantum theory have become the two pillars of modern physics, yielding many real-world applications, ranging from computer chips and GPS technology to the theory of black holes. Physics, through its language of mathematics, was forced to accept a new regime of reality that the Material consciousness could not, and still cannot, grasp. Quantum theory reveals a new worldview and an emerging new consciousness.

The collapse of classical physics and its mechanistic order was the end of an era in physics and demonstrates the limitation of the Material Mental consciousness. Now, a hundred years later, this limitation is blowing up on us in nearly every way. We see signs of failure in most of our civilizational systems, including the political, the economic, the educational, and the health care systems, not to mention the collapse of ecosystems from our ongoing assault on the biosphere.

---

80. Ibid.

These are all messages that the Material consciousness can no longer serve as our highest mode of operation. In physics something new and more complete did emerge from the limitations of classical physics, and I am suggesting that the same can happen for the limited Material paradigm. If humans are to continue flourishing on Earth, a new consciousness and culture must and can emerge. This is what Jean Gebser first envisioned in 1932, and what he spent most of his lifetime trying to convey to the world.

The Twentieth Century

If we were looking for a single year when a major shift in consciousness became apparent in the modern world, then 1905, the year Einstein appeared on the world stage, could reasonably be chosen. But this would be more symbolic than literal because the emergence of new structures of consciousness is not a sudden thing that happens during one particular year. More realistically, the evolution of consciousness occurs in fits and starts, on many fronts, usually over thousands of years, accompanied by retreats back to previous structures. We see such retreats happening today, for example, in the fear-based tribalism of white supremacists and Islamic jihadists.

Gebser's primary message was that a new consciousness structure was emerging in humans, and he dated its birth to the early twentieth century. However, it seems that this initial *irruption* (a Gebser term) between 1905 and 1930, which manifested in relativity and quantum physics, was stalled by the two great World Wars and the Great Depression in between. These wars were the culmination of centuries of toxic nationalism in Europe and became existential threats to global civilization, demanding the full engagement of the reliable Material-Mental regime. It was not time for a new reality.

The second world war ended with the detonation of the atomic bombs at Hiroshima and Nagasaki, a moment that completely changed the world. Science had harnessed the nuclear reaction, and it was now used for death and destruction in a weapon that was a million times more powerful than any previous weapon. After the war, a new generation was born in the shadow of the bomb, and as teenagers in the 1960s, they questioned and rejected nearly everything their parents held sacred. The cultural upheaval that ensued has been called the *counterculture*, and more recently *postmodernism*. The youth of the Sixties and Seventies visualized a much more connected world where peace and love could prevail, and they could not accept the material dreams and cultural silos of their parents.

Yes, the decade from 1965 to 1975 was filled with drugs, sex, and rock 'n' roll, but this period also saw the rise of the Civil Rights movements, environmentalism, the Women's movement, and a creative explosion in music. Pluralism, globalism, the freedom and empowerment of all people, and reconnection with nature were all signs of the new consciousness re-emerging after the great wars. Postmodernism was the first large-scale realization that humanity could not continue on the same path. The youth of this time — now called baby boomers — came into this world seeing things their parents did not see. They could see what was going wrong in the world, and they spoke out in protest. But they were too young and inexperienced to know how to build a better world, and the fundamental power structures and mindsets of the Material culture still prevailed. Today, many consider the counterculture a failed experiment, a naïve vision that stalled out. But I differ.

As a member of that generation who shared the dream of a better world and who fully participated in the counterculture, I consider this unfinished business — my business, our business. The project of the sixties, the dream of a new world, did not die, and it did not fail — it's just taking a lot longer than we expected. We tore down the old culture of our parents who were the "greatest generation," who lived through the Great Depression and the 2nd World War, but we could not solidify and sustain a shared vision of a new world. Humanity has been floating in limbo for the last 50 years, knowing that the old world is gone, but unsure if a new world is possible, or what it should look like, or how we will create it. Today, our future seems uncertain and dystopic, if not catastrophic. Many people have lost hope. But the seeds that were planted fifty years ago are still alive and growing.

We must now create a shared vision for the future of humans on this planet and a new consciousness that transcends the Material consciousness. It is now our mission — all of us who are capable — to recognize and support the shift from the Stage 4 Material consciousness into Stage 5 Planetary consciousness.

The Shift from Material to Planetary

Let us begin exploring this shift with the following chart that compares and contrasts Stage 4 and Stage 5 characteristics using simple descriptors. This is derived from Gebser and many others who have described the higher consciousness that awaits us.

5-1
EMERGING CULTURE AND CONSCIOUSNESS

Stage 4: Material/Mental → Shifts to →	Stage 5: Planetary
Centralized power	Empowerment of many
Dominator culture	Partnership culture
Top-down authority	Decision-making by consent
Coercion	Cooperation
Fragmentation	Wholeness
Separation	Integration
Categorization	Synthesis
Divided person	Whole person
Cognition	Verition
Thinking	Intuition
Reasoning and logic	Direct apprehension and epiphany
Head-centered thinking	Integration of head and heart
Ego-directive action	Collaborative-supportive action
Patriarchy	Co-equality
One perspective	Multiple perspectives
Win-lose dynamics	Win-win synergy
Exploitation	Service
Extract and dispose	Regenerate and re-use
Man *conquers* nature	We *are* Nature
Plunderers of the earth	Caretakers of Earth
National	Planetary
Either/or	Both/and
Classical Physics	Quantum Physics
Mechanistic universe	Organismic universe
Observer consciousness	Participatory consciousness
Causality and locality	Entanglement and non-locality
Objects and forces	Processes and relationships
Explicate order	Implicate order
3Dimensions of space	Higher dimensions
Vacuum and empty space	Quantum vacuum and the plenum

We should not necessarily think of the Stage 4 characteristics as bad and the Stage 5 characteristics as good. Nearly all stage theorists agree that new developmental stages *transcend and include* earlier stages in the same way quantum physics transcends and includes classical physics. As an example, we see "top-down authority" listed under Stage 4 as something we need to transcend. But in Stage 5, we can still *include* this in our repertoire of capabilities for appropriate situations. In a serious emergency, like a natural disaster, it is imperative that fast, effective decisions be made by a commanding authority and that these decisions (orders) be carried out quickly and exactly. There is no time to sit around and reach consensus! But when the disaster is over, we can put away that kind of authoritative, command-and-control decision-making process. The use of previous stages is situational.

The important point here is that in Stage 4 the characteristics listed above are the *only* way we can operate — we're limited to that. But in Stage 5, they become options, or tools, that can be used when appropriate, then put away. In Stage 5 we have access to all of the capacities and modes of the earlier stages, and more.

Yet there are some features of Stage 4 that we *should* consider to be harmful and should be left behind, such as the attitude that "Man conquers nature." Gebser emphasized that each stage eventually reaches a point of deficiency when dysfunction begins to set in. A textbook example of this is the Western world's wanton disregard for the natural environment, which began, at least, during the time of the Industrial Revolution. This is a deficient form of the Mental structure, and it is not only harmful, but it will be our undoing unless we can shift to a Stage 5 Planetary view of our relationship with nature.

Certainly, some of the characteristics listed in 5-1 under Stage 5 are not new. For example, *service* has been around for a long time, but it is not the dominant practice today, nor is it valued in today's Material culture. In a Stage 5 culture, service will be the norm, rather than the exception. Likewise, *intuition* is not new, and some people today recognize and use it effectively, even though hardcore materialists deny it. In Stage 5 consciousness, intuition is a well-developed capacity that is available *along with* thinking, reasoning, and logic. Intuition originates from the heart, and in Stage 5 the heart is activated and integrated with the brain (see chapter 6). Einstein was both a high-level mentalist *and* a gifted intuitive — a sure sign he had moved substantially into Stage 5 consciousness.

Stage 5 consciousness has been emerging in isolated pockets for a long time; it has been apparent in certain gifted individuals throughout history. Our work now is to help this become the primary mode of consciousness in humans and establish a Stage 5 culture on our planet. It is my hope that most readers will recognize and identify with many of the Stage 5 characteristics listed in 5-1 and realize that they are already manifesting. As more and more people recognize and grow into Planetary consciousness, a critical mass will be reached, and we will become a Planetary society where life can flourish.

A Speculation: Life in the Universe and the Technological Bottleneck

The following is a speculation on my part, and may or may not be true, but it will serve to transport us out of our tiny, Earth-based perspective, perhaps to see our situation today as part of something much larger.

This fantasy is a hypothetical answer to what has been called the *Fermi Paradox*. As the story goes, one day in 1950 the famous physicist Enrico Fermi was having lunch with some friends and asked, *Where is everybody?* He was asking about intelligent life in the universe and why we have never encountered it. This has been called a paradox because, on the one hand, we know that there are unimaginably huge numbers of star systems out there (stars with planets), and if just a very tiny fraction of these had planets suitable for life, the odds of intelligent life somewhere in our galaxy should be high[81]. On the other hand, paradoxically, we have never been contacted by intelligent life forms. Why not?

---

[81]. A recent study estimates that there should be at least 36 communicating civilizations in our galaxy. Westby, Tom, Conselice, Christopher J. *The Astrobiological Copernican Weak and Strong Limits for Intelligent Life.* The Astrophysical Journal (June 15, 2020).

Many speculations have been offered as answers to Fermi's question, so here are a few:

- We *are* alone. Earth *is* the only place where intelligent life exists.

- We *have* been contacted by intelligent aliens, and either (a) they are among us, or (b) the government has kept it a secret.

- The size of our galaxy (and the rest of the universe) is so great that communication among advanced civilizations is very unlikely. Our galaxy is about 100,000 light-years across, so a message traveling at light speed could take many tens of thousands of years to reach us. We have been listening intently since the 1970s (SETI research), but silence prevails.

- Intelligent beings *do* exist but have no interest in us, in the same way we have no interest in the affairs of worms.

- Intelligent beings *do* exist but choose to remain invisible to us.

There are many more speculative theories like these, but the last one provides the launching point for my own version. Let us suppose life not only exists in other places, but it is relatively common because, in fact, it is *inevitable* as a completely natural part of the evolution of the universe. If the conditions are favorable somewhere (liquid water is thought to be essential), life will take hold, and it will begin evolving. Even if this happens, many planets will experience catastrophes, like a major asteroid impact, that could eliminate life, or at least set it back. But if life has enough time, it *will* eventually evolve into something similar to the plants and animals on Earth, and eventually into beings with large brains and self-aware consciousness. In other words, what happened on Earth has happened in many other places.

This is what Young's theory of process is describing — nature evolves through the stages that he calls the seven Kingdoms: light, nuclear particles, atoms, molecules, plants, animals, and humans. These last three stages refer to *life*, and when we name them plants, animals, and humans, we are referring to how life evolved *on Earth*. At other locations where life emerges the same three stages would unfold, but not exactly as they did on Earth. The plants we have on Earth could more generally be described as organisms that can directly capture the energy of their star (photosynthesis), and therefore, they don't need to be mobile; animals could be generally described as organisms that must consume other organisms to metabolize their energy, and they, therefore, need to be mobile; and finally, humans are organisms with huge brains who are able to accumulate knowledge across generations and, thereby, surpass and dominate all other organisms on their planet. The important point is that if life can continue to evolve at any location in the universe, creatures analogous or equivalent to humans will inevitably emerge. In other words, intelligent beings with a high level of self-awareness are inevitable in the universe where conditions are favorable.

If so, we should expect intelligent organisms on other planets to follow a path similar to ours: they will start making and using tools from natural materials that are available to them; they will use these tools to hunt and process food; they will master fire; they will develop language; and they will build artificial structures to live in. They will become an "advanced" civilization. If this continues, they will eventually develop science, the very powerful tool for discovering the secrets of nature. Through the use of science, every intelligent species will eventually discover physics and chemistry because these are not human creations; these are descriptions of nature and how it works. They probably *won't* discover chess or basketball because these are strictly human creations, but they will eventually master the chemical reaction (such as the combustion of fuels and gun powder); and, inevitably, they will discover and harness the nuclear reaction, as humans did in the 1930s.

If a species is lucky enough to get this far (no extinction events), they will probably also live in a society that is fragmented into nation-like entities that compete with each other and use war to advance their own interests. That is, they will go through a Material stage of development. As weapons of war grow more powerful, the nuclear reaction will inevitably be used to make a weapon of mass destruction, and this now becomes existentially threatening. Things could play out differently in the details — perhaps altering their genome gets them in trouble instead of nuclear weapons, or perhaps they simply trash their environment and begin to suffocate in their own wastes. But eventually, the path of science and technology leads to the point where the species becomes a threat to itself because their technologies have become so powerful and dangerous. That is to say, every life form — if it can keep evolving — will eventually reach this point of existential crisis. Let's call it the *technological bottleneck*. Of course, this is exactly where humans are right now, and this is described clearly in Arthur Young's theory of process. This is the state of humans in Stage 4 of our evolution, where causality and materiality have fully emerged.

Let's further speculate that many species reaching this point (the technological bottleneck) do not make it through. In one way or another, they destroy themselves or stall out dystopically in their evolution. This is easy enough to imagine by looking at our own situation now. This scenario provides one answer to Fermi's Paradox: intelligent life-forms that develop science and technology tend to eliminate themselves because their power outruns their wisdom, so we never hear from them.

But let us further speculate that some species *do* survive this bottleneck by acquiring (just in time) the necessary wisdom and learn the right use of causality and technology. They make it through the technological bottleneck and are able to continue on their evolutionary path, acquiring unimaginable new domains of knowledge and moving into higher dimensionality. They successfully enter Stage 5 of their evolution.

But shouldn't we be hearing from them? The final part of my speculation is that we will also *not* be hearing from the beings who passed this evolutionary test because they know that humans, or any species, must accomplish this *by themselves*, or perish. They understand that this is Mother Nature's test, a filter that eliminates destructive, self-centered life-forms; therefore, they cannot and will not intervene in our affairs at this critical point. So, we won't be hearing from them. Only by passing this test *by themselves* can a species continue its evolution and become worthy of contact with the galactic community of other species who made it through the technological bottleneck by achieving higher consciousness.

So, to answer Fermi, there will be no contact, only silence, until we have evolved into the next stage of our development, where we can use science and technology for the benefit of all. Only then will we make contact with other life forms who have also transcended the destructive grip of Stage 4 and evolved further. But until then, there will only be silence.

End of speculation. Of course, it may be some time, if ever, before we know whether we are on a standard evolutionary path that many other life forms throughout the universe have been on or if we are alone in our predicament. Either way, it is clear that the power of our technologies to do harm now far exceeds our level of wisdom. Acquiring the wisdom we need to survive our technological bottleneck can only come from the evolution of consciousness. We are at the end of a major era in our story, and a new era is possible in which life can thrive on this planet and humans can continue to evolve.

CHAPTER 6

Seven Markers

Human consciousness has been evolving for the last three million years, and it still is today. Our exterior physical form has not changed visibly for hundreds of thousands of years, but what we are *inside* certainly has. The evolution of consciousness is what sets humans apart from all other living things, and it is the hope for a better future. We have seen that there have been four main chapters in our story — the Mimetic, Magical, Mythical, and Material — that reflect the evolution of consciousness and culture.

We have explored our present culture and consciousness, the Material stage, and seen why it is not sustainable, and now, we look ahead to our next stage of evolution and the emerging new structure of consciousness. This has been foretold and described by Jean Gebser, Sri Aurobindo, Pierre Teilhard de Chardin, Arthur Young, and many other visionaries, and it has been called by many names. I have called it the *Planetary consciousness* because we are awakening to a new relationship with the living planetary system, Gaia, and a new relationship with all of humanity. We are becoming *Planetary humans*, not merely Russians or South Africans, Jews or Christians, blacks or whites.

This chapter is devoted to illuminating the Planetary consciousness to the extent that it is possible in words. What follows is a collection of ideas and metaphors from many sources about the nature of "higher consciousness," as sages, philosophers, and spiritual teachers have spoken of. These ideas reflect what I have discovered and found meaningful over fifty years of searching and researching. I have organized them into seven general themes, or *markers*. These are broad descriptors, or landmarks, or look-fors, but not distinct categories with clean boundaries; they overlap and interrelate. I did not call these THE Seven Markers, just *seven markers*, because they do not represent the final word or the ultimate truth. We could expand these into more than seven or collapse them into fewer, and I have undoubtedly missed a few things. This is intended to be a starting point that stimulates thought and discussion and that creates a framework and language for understanding higher consciousness. My hope is that others will add to this framework and improve it, so that we can bring the knowledge of evolving consciousness to everyone.

I hope you will recognize many of these features of Planetary consciousness because they are already emerging in *you*, and in many of us; they have been seen for thousands of years in a few individuals who were way ahead of their time. It is now time for large numbers of people to recognize and awaken our Planetary consciousness. This is the hope for the future of humanity.

Here are the names I will use for the Seven Markers that we will explore in this chapter:

1. **Awakening**
2. **Connectedness**
3. **Meta-perspective**
4. **Softening of the Ego**
5. **Wholeness**
6. **Heart Opening**
7. **The Evolutionary Worldview**

1. Awakening

When we awaken from deep sleep or a dream, we return to our normal conscious awareness of the world and who we are. This happens to all of us at least once a day. But awakenings of another kind happen when we are fully conscious and aware in everyday life. A *conscious awakening* is the experience of suddenly gaining a new understanding or meaning, an insight, or perhaps a revelation. We may experience it as an "aha moment" or when "the light goes on" or as "putting two plus two together." An awakening is a discrete increase in *knowledge* (see Side Bar) that gives us a more complete and more accurate understanding of the world or ourselves.

Science cannot yet explain what's happening in the brain when we have an awakening, but we may presume that new connections, new circuits, or new resonant states are being established. When we have an awakening, we increase our conscious capacity because new understandings and meanings have been created: this is an increase in knowledge and an incremental intensification of our consciousness. The awakening, then, is the increment of evolving consciousness, a step forward in personal evolution. The awakening is the means by which consciousness gradually evolves.

An awakening is often brought about by the arrival of new information, even one tiny piece of information, making separate pieces fit together into a whole picture that suddenly becomes meaningful. But awakenings can also be quite spontaneous, seemingly for no reason, at times when the mind is relaxed. Fragments of information are integrated into a whole that takes on a new meaning. This is the process of turning information into knowledge.

Typical awakenings open us up, in small steps, to seeing a more complete and truer picture of the world. They remove our blinders, dissolve our preconceptions, and increase our knowledge of reality, usually in tolerable doses. But sometimes awakenings can be unwelcome or traumatic, forced on us as a "wake-up call." A brush with cancer, a near-death experience, the loss of a loved one — all of these shatter our assumptions and wake us up to what is real and meaningful in life.

How is awakening different from learning? Unfortunately, learning, as most of us experienced it in school, is often shallow and centered on memorization and recall of meaningless information. This level of learning should more accurately be called conditioning. With ample rewards and punishments, students are trained to store information and recall it when asked, but there is no *meaning* in this process. The truly great teachers do in fact create awakenings in their students. Students gain knowledge from great teachers, and their understanding of the world increases. Authentic learning *is* awakening.

Sometimes awakenings can be spiritual in nature, or what Abraham Maslow called *peak experiences*. These non-ordinary experiences of reality have been reported by mystics throughout the ages, but also by ordinary people appreciating nature or performing at a high level or simply at odd moments when thinking has been calmed down.

> ### DATA, INFORMATION AND KNOWLEDGE
>
> DATA are the result of an observation or measurement. The temperature of the air around you can be measured and turned into a datum (singular). If I take a digital photo of a tree, the camera collects a large amount of data from the light being reflected by each part of the tree. These data are nothing more than zeros and ones in the memory of the camera. Taken alone, data are meaningless.
>
> Data, such as strings of zeros and ones, become INFORMATION when organized into a bigger structure. The data captured by the camera become information when a computer translates the binary code (zeros and ones) into pixels with locations, creating an image of the tree, perhaps as a jpeg file. This information, the digital photo, lives in computer memory, or can be displayed, but it is still meaningless by itself.
>
> KNOWLEDGE is generated when information becomes *meaningful*, and this requires a human. When I look at the photo of the tree, my consciousness receives that information and puts it together with my existing knowledge, and that information can become meaningful to me. Perhaps the tree in the photo takes on the meaning of "blue spruce", a kind of tree I find beautiful. The *data* collected by the camera that became *information* as the photo now means something to a human. Knowledge has meaning, while data and information mean nothing without a human. *Meaning* is a human subjective experience, that requires an owner – a self. Machines such as cameras and computers can deal with data and information, but only human consciousness can acquire knowledge.

Nearly all people who have had intense peak experiences report feelings of oneness with the cosmos, intense aliveness, a unity of mind and body, a profound sense of connectedness and non-duality, and heightened alertness and clarity. These are usually life-changing experiences.

Our day-to-day awakenings are usually small and unremarkable, but there are many reports throughout history of people having *big* awakenings, referred to by many names: *satori, samadhi, illumination, gnosis, epiphany, moksha, kensho, transcendence,* or, the mother of all awakenings, *enlightenment* as experienced by the Buddha. But awakenings need not be mysterious or esoteric — we don't have to burn incense, wear robes, chant, or go to church regularly to have access to them. Awakenings are natural and available to all of us. Our challenge is to be *open* to them.

The problem for most of us is that the natural possibility for conscious awakening gets closed off and disrupted by things like fear, anxiety, rigidity, excessive thinking, cultural conditioning, and fragmented attention. Many people today are paralyzed by fear promoted through the media, because fear persuades and sells. And fragmented attention is epidemic, driven by the constant electronic interruptions of beeps, dings, rings, notifications, alarms, and updates. We must learn to put our intrusive technologies aside when they are not serving a clear purpose and reclaim our own inner space where awakenings can happen naturally.

Cultural conditioning is a powerful ubiquitous influence that puts us in boxes with artificial walls, which are defined by other people — parents, teachers, friends, bosses, religious figures, advertisers, pop icons, and more. We end up trying to be like someone else, rather than consciously and freely choosing what we want to be and to do and what we aspire to. If we want to be more open to awakenings and find our own unique gifts, we need to examine and question our assumptions and discover what is conditioning and what is our own inner wisdom. Conditioning limits us while awakening expands us.

Awakenings are also stifled by the constant chatter of the "monkey mind" and the fragmented attention that restlessly jumps from one thing to the next. Modern humans operating primarily in the Mental consciousness are dogged by obsessive thinking and a scattered mind. Our internal dialog can keep repeating stories, sometimes ones that are not even true, creating fears, anxieties, and misunderstandings. This drowns out our deeper wisdom and creativity. But we can settle down our monkey mind by cultivating more stable attention. This allows *conscious* choices and actions, sometimes called *mindfulness*, instead of conditioned choices and actions. Mindfulness means choosing and acting with full awareness.

How does one cultivate stable attention and mindfulness? Something that can help is a regular *awareness practice*. There are many awareness practices, including many types of meditation, and I have suggested a basic breath awareness exercise below for those new to this. Allow just ten or twenty minutes each day to close your eyes and sit quietly, to stop doing and just be. This helps stabilize the attention, instead of having it jump around uncontrollably. It also helps quiet the obsessive thinking that dominates our waking life, creating space for awakenings to naturally emerge instead of being drowned out by mental chatter. With some practice, you'll find yourself becoming better at focusing your attention on one thing, like truly listening to another person. Awareness practices are inner exercises that help us get healthy inside, just like exercising the body. This allows the natural process of evolution to take place within us because we are more receptive to awakening.

BASIC BREATH AWARENESS PRACTICE

1. Sit comfortably in a quiet place, then close your eyes and relax.

2. Bring your attention to your breath - feel it moving in and out through your nose as it fills and empties your lungs. Feel the air moving past the very tip of your nostrils.

3. Try to keep your attention on your breathing for ten breaths, in and out. This is not easy and you will probably drift into thoughts after a few breaths. Not a problem! Let those thoughts slip by like clouds passing in the sky, and gently return your attention to the breath. Start out doing this for ten minutes, then gradually increase to twenty as you are ready. That's it.

---

2. Connectedness

Connectedness is a primary feature of higher consciousness that is widely recognized across wisdom traditions and indigenous cultures, but most of us today are oblivious to it. Nature-based cultures necessarily understand the ecological web of connectedness within the natural world, but most of us who live in material civilization are so out of touch with nature that we have lost our sense of connectedness. The Material regime of consciousness blinds us to connectedness by generating the reality we know so well, one consisting of lifeless objects in space. David Bohm calls this exterior world of objective reality the *explicate order*. But, according to Bohm, underlying this material veneer is the *implicate order*, a higher dimensional, non-material, invisible medium that connects all things manifesting in our material world. Let's explore what has been said about the implicate order of connectedness.

Traditional Views. Indigenous South Africans use the word *Ubuntu* to express the connectedness of humanity and the community of all people. Ubuntu is sometimes translated as "I am because we are." It expresses the core of our human-ness that began with our ancestors three million years ago — our existence as socially interconnected mimetic creatures co-creating culture. Ubuntu means that we are not separate things, but we exist together. It reminds us of our connectedness.

Aboriginal elder Bob Randall explains indigenous Australians' view:

> For the Yankunytjatjara Aboriginal people from northwest South Australia the law of Kanyini implies that everybody is responsible for each other. It is a principle of connectedness that underpins Aboriginal life. And because of connection, Kanyini teaches to look away from oneself and towards community: we practice Kanyini by learning to restrict the 'mine-ness,' and to develop a strong sense of 'ours-ness.'[82]

Interbeing is another word that conveys connectedness. It was originated by Thich Nhat Hanh, the Vietnamese Zen Buddhist monk and renowned peace activist. Interbeing is based in the Buddhist principle of *dependent origin*, which holds that all phenomena are interdependent; in other words everything is connected. "To be" or "being" conveys solitary existence in separateness, but "to inter-be," or "interbeing," is to exist in connectedness.

---

82. From *Songman: The Story of an Aboriginal Elder*. Australian Broadcasting (2003).

Author and peace activist Scott Brown explains interbeing in ecological terms:

> From the bacteria in our guts, the decomposers in and of the soil, the insects that provide pollination, the plants and animals we eat, to the phytoplankton, trees, and plants that absorb carbon and create oxygen — we are all in it together, interconnected in a web of relationship and interdependence. Interbeing is a basic truth of our existence.[83]

<u>The Fallacy of Empty Space</u>. In the material world, separation and empty space go hand-in-hand: empty space is what makes separation possible. It *defines* objects. It is the *objects* that are the stars of the show in the material world, but they could not exist without empty space. However, there is mounting evidence from science showing that the notion of empty space is not intrinsically valid — it's an approximation that works within the limited reality of the material world, the explicate order. Quantum theory now demonstrates that the connectedness of the universe and all living things comes from the hidden properties of space, which is not "empty." The equations of quantum field theory reveal that the vacuum of space contains a *zero-point energy*, a background *something* that is non-material and fills "empty space." When the zero-point energy is calculated, it adds up to an immense sea of energy contained in vacuum.[84] So much for empty space.

Astrophysicists also concur that there is much more to the universe than just matter and empty space. They now believe that all the matter and energy in the universe that we can currently detect is only about 4 percent of the contents of the universe. Another 70 percent is "dark energy" whose nature is unknown.[85] In other words, science does not yet know what 70 percent of the universe is comprised of because it doesn't show up on any of our instruments, yet there are many hints that *something* is there.

---

[83]. From *Active Peace: A Mindful Path to a Non-Violent World*. Collins Foundation Press (2016).
[84]. ~ 0.6 eV/cm^3 according to cosmologist Sean Carrol.
[85]. The other 26% is thought to be *dark matter* that is invisible (it does not emit any form of light we can detect) but is clearly evidenced by its gravitational effects on matter that we *can* see.

In the most recent mathematical models of the universe, dark energy was invented to account for the accelerating expansion of the universe that has recently been observed, but no one knows what it actually is. Einstein proposed his famous *cosmological constant* in 1916 to keep the universe from collapsing under gravity, but he later rejected it as his "greatest blunder." Now we've resurrected it as dark energy, so it seems Einstein was right after all. Perhaps dark energy and zero-point energy are both manifestations of the same higher-dimensional field of the implicate order. From both quantum physics and astrophysics we arrive at a similar conclusion about the world, about the nature of reality: there is a very big *something* behind the scenes, behind the façade of the material world and its alleged empty space.

This is an apt description of our own view of the world: what we call *reality* is a tiny fraction of what's actually there. The world of material objects that we see from our limited consciousness is a highly-filtered veneer over a deeper underlying reality. As humans continue to awaken into higher consciousness, we will be more able to know the implicate order of reality. The great frontier for humans is *inner* space, not *outer* space.

Non-locality and Entanglement. Quantum theory has led to a radical shift from understanding space as that which *separates* everything, to that which *connects* everything. One of the predictions of quantum theory that disturbed Einstein was the phenomenon of *non-local* connection across space, popularly called *entanglement*. So, what is non-locality?

In our material world, things happen through *local* action. If you want to move a big rock, you must be nearby to give it a push, to apply the force that causes it to move. But that can't happen if the rock is in California and you're in New York. You have to be *local*. You *could* send a message from New York to a friend in California and ask her to push the rock, or you could operate a bulldozer remotely from New York. But these tricks would still constitute local action because the actual push is local. That rock is not going to move unless something nearby pushes on it and *causes* it to move. This is *causality*, or cause and effect. Locality and causality are the foundations of Stage 4 Material reality — the world of separate objects in space. It's common sense. Things happen for a reason. There is a cause for every effect, and the cause must be local. If we see an object floating in space for no apparent reason, we know that if we investigate further, we will find a cause, such as hidden supporting strings or levitation in a magnetic field. Afterall, this *is* the real world, right?

Yes, this is the *material* world we know, the explicate order of reality that is governed by causality and locality. But physicists have now been forced to accept that non-local action, or non-locality, is also part of the real world. Non-locality means, in principle, that there *could* be a way to move the rock in California, nearly instantaneously, through actions taken in New York, or from across the galaxy for that matter.

Non-locality has been demonstrated at the atomic level in experiments since the 1970s. The simplest kind of experiment takes two electrons that have "cozied up" by being near each other, perhaps by occupying the same atom. They share the same quantum state — they are *entangled*. If we now separate them to a great distance, then disturb one of the pair (its spin is altered), the other one responds almost instantly, faster than a message could be sent through 3D space at the speed of light. The two separated particles seem to act as one. They have somehow remained connected, or entangled, "behind" the framework of 3D space, interacting *superluminally* (faster than the speed of light).

More recent experiments have demonstrated quantum entanglement in atoms and in larger systems of molecules. It has now been shown by several teams of researchers that the light-harvesting mechanism of plants, photosynthesis, relies on the quantum entanglement of biomolecules to transfer energy almost instantaneously.[86] Another team recently announced that they had observed an entire bacterium entangled with light.[87]

---

86. Engel, G.S., et al. *Evidence for wavelike energy transfer through quantum coherence in photosynthetic complexes.* Nature, 446, 782-786 (2007). Collini, Elisabetta, et al. *Coherently wired light-harvesting in photosynthetic marine algae at ambient temperature.* Nature, 463, 644-647 (2010).

87. Marletto, Chara, et al. *Entanglement between living bacteria and quantized light witnessed by Rabi splitting.* Journal of Physics Communications. (10 October 2018).

Entanglement is now firmly established and at ever-larger scales, supporting the view that all living things and complex organs, like the brain, are massively entangled systems that function as connected wholes, not merely as classical systems of parts that operate through chemical reactions and electrical firings, like a machine. Penrose and Hameroff (Chapter 2) have proposed that the firing of individual neurons and global neuronal activity are surface mechanisms fed from below by the information originating in the vast microtubule system in every one of our cells, where quantum mechanics and entanglement dominate.

If non-local connectivity is real, as experiments in physics now show, this opens the door to explaining many phenomena that have been dismissed by science, such as non-local healing, telepathy, and remote viewing. We can understand why Einstein was so uncomfortable with all this, calling non-locality and entanglement "spooky action at a distance." Non-local connectedness seems impossible to anyone with common sense and rationality. Yet the evidence is mounting in support of non-material connective fields, transcending space, that we cannot detect with our existing technology.

Equally startling is non-locality in *time*. The Material consciousness restricts us to a linear sense of time moving forward inexorably, with each moment flowing from the previous moment — temporal locality. Past moments and events are accessible to us only as memories that are assumed to be stored somewhere in the brain and can be called up, in much the same way computers store information in specific locations to be retrieved when needed. However, neuroscientists have had difficulty locating memory in specific brain areas or finding the mechanisms of storage and retrieval.

Karl Pribram has proposed that memory and other brain functions are spread out holographically over large areas of the brain,[88] while others have suggested that memories are stored across the entire body, perhaps in connective tissue. But even more radical is the idea that the past is recorded *outside* of the brain and body in non-material informational fields. This

---

88. *Brain and Perception.* Lawrence Eribaum Associates (1991).

would mean that the past is not gone but is still there and accessible to us. Non-locality in time means that past, present, and future are not separated by linear time but exist together in a vast, non-material, informational field (see Laszlo and the *Akashic Field* below). This sounds just like Gebser's notion of *time freedom* in the Integral consciousness in which past, present, and future are no longer separate. Yet for now, it seems that our brains are hard-wired for locality in both time and space, which creates our world of separate objects moving forward through time.

While mainstream science still largely holds that consciousness in all of its aspects is produced entirely *inside* the physical brain, it now seems possible that the brain also acts like a receiver that can tune into informational fields, like a radio receiver that tunes into the sea of radio waves filling space. If we heard a symphony orchestra playing though a radio, we would not look for the musical instruments *inside* the radio, yet we continue to look for consciousness *inside* the brain.

The Akashic Field. Systems scientist Ervin Laszlo has articulated the nature of non-local connective fields perhaps better than anyone today. He describes an underlying higher-dimensional medium, the *unified field*, a universal non-material substrate that can manifest as the basic fields of physics — the electromagnetic, the gravitational, the strong nuclear, and the weak nuclear — and also the quantum fields that give rise to sub-atomic particles.

Another aspect of the unified field is the *Akashic field*, an idea introduced thousands of years ago in the Vedas. The Sanskrit word *Akasha* loosely translates to sky or space, so the Akashic field is the "space field." Some ancient traditions refer to the *Akashic record,* a record of all that has ever happened which is stored in this space field. In the early 1900s Rudolf Steiner brought the concept of the Akashic field to the western world, describing it as a field of information, but it was largely forgotten, except in the most esoteric circles, until Laszlo resurrected it in two recent books.[89] According to Laszlo,

---

[89]. *Science and the Akashic Field.* Inner Traditions (2004), and *The Akashic Experience.* Inner Traditions (2009).

The unified field is a space-filling medium that underlies the manifest things and processes of the universe… It carries the universal fields: the electromagnetic, the gravitational, and the strong and weak nuclear fields. It carries the field of zero-point energies. And it's also the element of the cosmos that records, conserves, and conveys information. In the latter guise it's the Akashic field, the rediscovered ancient concept of Akasha.

The unified field contains information we can access, some people more easily than others. The genius discoverer Nikola Tesla (1856-1943) described an "original medium that fills space"[90] from which he was able to receive information. Out of "nowhere" he saw whole pictures of inventions and phenomena of nature. He saw, in a single gestalt, the electrical power system based on alternating current that we now call *the grid*, and he literally downloaded into his own mind the complete blueprints for the induction motor/generator during a brief episode in the streets of Prague with his brother. Although Tesla's gifts were extraordinary, all of us have probably accessed information from higher-dimensional fields in the form of hunches, intuitions, or just *knowing without thinking*.

A shared informational field throws into question the whole notion of individual people making new discoveries and having "original" ideas. Are these actually new ideas that originated completely from within the brain of one person, or does the discoverer receive information from "out there" by acting like an antenna tuned-in to these informational fields? In many cases, new discoveries are made almost simultaneously by people working independently. The person who publishes first ends up getting the credit, but perhaps the co-discoverers were sourcing from the same informational field. The existence of shared information fields also suggests an explanation for many parallel events in history that occurred at widely separated locations; for example, the early megalithic structures that showed up on different continents at similar times and the parallel appearances of civilization in the Middle East, Africa, Asia, and the Americas; or, the almost iden-

---

[90]. From Ervin Laszlo in *Science and the Akashic Field*.

tical tool cultures developed by Neanderthals and archaic humans while living on separate continents.

Ancient wisdom and modern science both point to the connectedness of the world we see as separate objects. The problem for most of us is that we cannot "see" connectedness in our daily affairs in the material world, so we continue to believe in separateness. This makes us believe we can harm other living things and damage the natural environment without any consequences for us, that our living planet is inert and dead and there for us to exploit. But we are evolving out of the worldview of separate objects and into the world of connectedness and relationship — Planetary consciousness. We can become more receptive to our connectedness by listening deeply to nature and reawakening our sensitivities, to see with more than just our eyes. Once we start to realize that we are connected to each other and to all of life, it changes everything about how we choose to act.

3. Meta-perspective

This marker comes straight from Gebser, but I propose to call it *meta-perspective* instead of Gebser's term, the *aperspectival* — it's a little easier to say. *Aperspectival* means *without a single perspective*, and Gebser considered this to be the primary characteristic of the Integral consciousness, what we are calling Planetary consciousness. Meta-perspective means to have *perspective on your perspective*; this means we have stepped back enough in our view of the world to recognize our *own* perspective. It means we have acquired a much bigger view of the world and ourselves.

Gebser was not the first to recognize this larger view of the world. Marcus Aurelius, the second-century Roman philosopher-emperor, often practiced the "view from above" that he attributed to Plato:

> This is a fine saying of Plato: that he who is discoursing about men should look also at earthly things as if he viewed them from some higher place.[91]

---

[91]. Aurelius, Marcus. *Meditations*. Book 7, number 48.

The seventeenth-century philosopher Baruch Spinoza originated the phrase *sub specie aeternitatis,* which means *from the perspective of the eternal.* This is the timeless, universal view of the world. In modern times, the most famous statement of the view from above (literally) is from astronaut Edgar Mitchell, the sixth man to walk on our Moon:

> In outer space you develop an instant global consciousness, a people orientation, an intense dissatisfaction with the state of the world, and a compulsion to do something about it. From out there on the moon, international politics look so petty. You want to grab a politician by the scruff of the neck and drag him a quarter of a million miles out and say, 'Look at that, you son of a bitch.'[92]

According to Gebser, the Mental consciousness is *perspectival* — it has a single perspective. We are all so familiar with this it's easy to overlook, because the single perspective we are talking about is *one's own* perspective. What other perspective can I have? I can only see through *my* eyes, and hear with *my* ears, think *my* thoughts, and have *my* opinions, right? Yes, in the Mental consciousness. But in the *aperspectival* consciousness — the meta-perspective — one leaves the self-centered perspective and can simultaneously hold multiple perspectives.

Mental consciousness generates the Cartesian duality — the subjective-objective separation, the inner and outer of the modern human experience. The subjective is *me* centered — my thoughts, my feelings, my perspective on the world biased heavily by my needs and my desires — while the objective is the cleanly separate world "out there" where I must do battle to survive and possibly thrive. *Me versus the world* is the fundamental tension and loneliness of life in the Mental regime of the modern world. But the meta-perspective blurs this separation — the self and self-interest become less central and more integrated with the world, including the interests of others. Having perspective on our perspective means we are transcending the limitations of self-centeredness and duality. The subjective and objective become less distinct; the boundaries between self and other are

---

92. *People Magazine,* April 8, 1974.

softened, and connectedness becomes more apparent. As we acquire the meta-perspective, we can zoom out and hold multiple perspectives; we can consciously change perspectives and more easily put ourselves in someone else's shoes.

The world begins to look different as we transcend the single perspective. We become less invested in maintaining and defending the hard boundaries of the Stage 4 self as we move into Stage 5 selfhood. This letting go of the me-centered perspective is what Gebser called *ego transparency*, which we will explore in the next section, *Softening of the Ego*. Meta-perspective always brings a bigger picture and a more complete understanding of the world. We can still have our own individual needs and self-interests and our own unique perspective, but these can exist simultaneously within the more-encompassing meta-perspective.

There are many influences in our environment that narrow our perspective, resulting in fixed opinions, tunnel vision, and getting stuck in a box. We are barraged with powerful messages from our upbringing, from our schooling and religion, from our social groups, and from seductive images delivered through our electronic screens that show us what we should look like and be like and tell us how we should think and feel, how we should vote, and, of course, what we should buy. Screens are direct portals into our brains, and at least some part of the content we are consuming is designed to make us think something and do something we wouldn't otherwise — mostly spend money.

Ultimately, it is the bigger worldview of the meta-perspective that allows us to see all these forms of conditioning. We must try to cultivate more openness to other perspectives by stepping back and widening our vision, softening our boundaries, and letting go of our categories. When we have the ability to zoom out from the self-centered world of the single-perspective, we see a bigger picture of everything, including ourselves. We become more able to witness our own choices and actions, as though viewing them from the perspective of others.

According to Gebser, Pablo Picasso (who he knew) was trying to convey the aperspectival consciousness in his paintings from the Cubist period around 1910. We see strange figures looking forward frontally, yet there is a side view of a nose — a profile. These paintings show two different perspectives simultaneously, implying a transcendence of time and space. In the world of 3D space and linear time — the Mental consciousness — we can take a front view, *or* we can take a side view, but they must be separated in space and time: we would have to move from the front to the side, a movement in space, and this would take some amount of time. But in the full aperspectival consciousness, one becomes free of time and 3D space, so these two perspectives are no longer separate — they become one. Admittedly, this is hard for most of us to comprehend!

More pragmatically, the meta-perspective naturally gives rise to empathy and compassion. These are two different things, but both are accompanied by a loss of self-centeredness and the single perspective. Empathy is the ability to truly recognize and even experience someone else's feelings and emotions. It is a *felt* experience. But with compassion you not only feel someone else's suffering, but you want to alleviate it. You take action. People who are strongly empathic can be overwhelmed, drained, and burned out by their heightened sensitivity, but compassion, with its component of action, can be regenerative. The compassionate caregivers of the world, like Mother Teresa, embody the higher consciousness of the meta-perspective.

4. Softening of the Ego

We return once again to the greatest of enigmas, the *self*. For thousands of years scholars and sages alike have tried to discover what it is, but from Buddhism to neuroscience there is still no consensus explanation for the self. Despite this difficulty, there are some things we *can* say.

First, there is widespread agreement that the self is not a thing we can find or a structure in the brain. It is best understood as a *process*. Evan Thompson expresses this beautifully:

When I say that the self is not a thing but a process what I mean is that the self is a process of "I-ing," a process that enacts an "I" and in which the "I" is no different from the I-ing process itself, rather like the way dancing is a process that enacts a dance and in which the dance is no different from the dancing.[93]

Secondly, there are certain misconceptions about the self that arise from Eastern philosophies misinterpreted in the West. One is that the self is an illusion — it is not real. This originates in the idealist Yogācāra school of Mahāyāna Buddhism. However, to say the self is an illusion is like saying *the world* is an illusion. This *can* be argued philosophically, but it really does us no good at all. We may have *delusions* about who we really are, and we can create *illusory* personas, either consciously, like an actor, or unconsciously. But none of this supports the claim that the self is not real.

Another misinterpretation of Buddhism is that the self, often referred to in Buddhist translations as "the ego," should be *eradicated* — it is the enemy. However, this is a very dangerous proposition because we must have a self of some kind to participate in the social milieu of culture. We need a self in order to operate in the world, so we shouldn't strive to eradicate it. What Buddhists are trying to say is that we need to evolve beyond a certain *type* of self that has been called the *ego*. Let us agree, then, that the self is both *real* and *necessary*, and shift our attention to the *evolution of the self*.

The word "ego" must be clarified before we go further. It was introduced in the early 1900s by Sigmund Freud in his theory of the psyche (what he called "personality"). He proposed that the psyche is balanced in three parts: the *ego*, the *id*, and the *superego*. In Freud's theory, the *ego* is the "executive self" that navigates the real world; the *id* is the more instinctive primal drive; and the *superego* is composed of the intellectual, spiritual, and higher sensibilities of personality. Though Freud was very influential, this three-part model of the human psyche is not the dominant theory of mind in psychology today (and some say there is none). Since Freud's time, the word *ego* has taken on somewhat different connotations, as expressed in the

---

93. From *Waking, Dreaming, Being*. Columbia University Press (2015).

insult, "Bob has a big ego." This means that Bob is very self-centered, if not narcissistic. Given the possible confusion about what this word means, I will avoid using the term *ego*, and instead use *egoic self*. The egoic self is a *type* of self, a *stage* in the evolution of the self, but it is not THE self. The egoic self is not the only form of selfhood, but it is the one that predominates today, so we are all very familiar with it.

Jean Gebser identified the egoic self as an evolutionary stage that was intrinsic to the Mental structure of consciousness and its *perspectival* view of the world. The perspectival consciousness sees the world from a single perspective — *mine*. The perspectival worldview by its very nature puts ME at the center of the universe, and the egoic self, by its very nature asks, what's in it for ME? Some might argue that such a mindset is optimal for Darwinian survival, to look after yourself above all else. However, humans evolved in the universe of culture, in which pure self-centered *me*-ness does not work as well as cooperative *we*-ness. The ability of our ancestors to collaborate, share, and problem-solve together provided a significant competitive advantage, and that is still true today. Mind-sharing culture has been our hallmark from the beginning. Yet this stage of selfhood, the egoic self, persists widely today in mainstream culture. Perhaps it is not serving us so well.

When the egoic self feels it matters more than anyone else, we call it *selfishness*, a condition that is considered "normal" but not desirable. More extreme than this is *narcissism*, a recognized personality disorder[94] characterized by exaggerated feelings of self-importance, an excessive need for admiration, and a lack of empathy toward other people. Going even further down this spectrum to its extreme is a philosophical position known as *metaphysical solipsism*, which holds that the *only* thing that exists is ME. That is, the external world, including other people, has no independent existence. In popular language, *you are a figment of my imagination*. Of course, no one seriously takes this position except philosophers playing

---

94. *Diagnostic and Statistical Manual of Mental Disorders, Version 5*. American Psychiatric Association (2013).

devil's advocate, and, hopefully, most of us are operating near the other end of this spectrum.

The stereotypical, over-developed egoic self, characteristic of the Material stage of consciousness, is easy to spot from the incessant self-promotion that exudes. *It's all about me* is the subtext of everything. This is captured nicely by the hyper-narcissistic alien named Q, speaking to Jean Luc Picard, Captain of the Starship Enterprise:[95]

> I've said enough about me. Now, what do *you* think about me?

For the inflated egoic self, every aspect of social life falls somewhere on a ladder of status — other people are either above me or below me on that ladder. The egoic self wants to be smarter, more popular, faster, more beautiful, richer, more powerful, more charming, more successful, and just flat out better than you. And one good way to look better, to elevate yourself, is to push others down. For the over-developed egoic self, all that matters is what can be gained for *me* and how my status can be elevated compared to others. Of course, not everyone is like this today! I am exaggerating to make a point.

The world we live in is still dominated by the egoic form of self. It is encouraged and rewarded in many ways. The fastest way to be "successful" in the Material culture — to acquire money, power, and attention — seems to be through wielding a strong, self-promoting egoic self. The Trump model captures this well. But there is a new way to be in the world, a new form of selfhood characteristic of the next stage of human evolution.

Our self is a part of our consciousness (an important part according to Antonio Damasio), and because consciousness is evolving, *the self is evolving*. As culture and consciousness have evolved through four broad stages, so has the self. As humans transition into the 5th stage of our evolution, the self is similarly evolving from an egoic, self-centered form characteristic of Stage 4 *Materialism* into a new form, the Stage 5 self.

---

95. From *Star Trek: The Next Generation*.

Gebser used the term *ego transparency* to describe the softening of the egoic self. Metaphorically, the egoic self is protected by hard walls; it presents as coercive and "in your face"; it uses force to make things happen. But the Stage 5 self is softer, more cooperative, and more transparent. It exists in *relationship* with others. Rather than being less powerful, it is in fact more powerful because it understands the power of collaboration and of empowering others. It also understands the right use of power to promote the well-being of others, which becomes more obvious when the blinders of self-promotion fall away.

Abraham Maslow's well-known hierarchy of needs, shown in Figure 6-1, places *self-actualization* at the highest level. In this hierarchy, the higher needs are not accessible until lower needs are satisfied. If you are hungry, cold, and afraid, you can't be concerned with self-esteem, art, or philosophy. The different levels also correspond roughly to different life stages. The needs at the bottom of the pyramid are predominant in infancy and early childhood while the needs for belonging and self-esteem emerge in later childhood and early adulthood, and the need for self-actualization appears in mature adulthood. Self-actualizers have strong identities — they know who they are, what they want, where they are going, and what they are good at. They are strong selves. Self-actualization is the acquisition of the fully individuated, *healthy* egoic self.

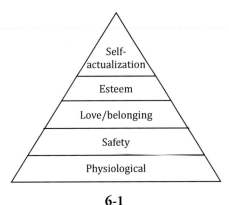

6-1
Maslow's Original Hierarchy of Needs

Maslow extended this theory near the end of his life, and in his later thinking he recognized a higher level of development that he called *self-transcendence*. In his last published work, he writes:

> Transcendence refers to the very highest and most inclusive or holistic levels of human consciousness, behaving and relating as ends

rather than means to oneself, to significant others, to human beings in general, to other species, to nature, and to the cosmos.[96]

Maslow further describes people who have self-transcended in the following terms:

- There is more and easier transcendence of the ego, the Self, the identity.
- There is a transcendence of competitiveness, of zero-sum, of win-lose gamesmanship.
- They have transcendent experiences and illuminations that bring clearer vision … of the ideal … of what ought to be, what actually could be.
- They are more apt to regard themselves as carriers of talent, instruments of the transpersonal, temporary custodians so to speak of a greater intelligence or skill or leadership or efficiency.
- They are more holistic about the world than are the "healthy" or practical self-actualizers … and such concepts as the "national interest" or "the religion of my fathers" or "different grades of people or of IQ" either cease to exist or are easily transcended.
- They are far more apt to be innovators, discoverers of the new, than are the healthy self-actualizers.
- There is a total wholehearted and unconflicted love, acceptance … rather than the more usual mixture of love and hate that passes for "love" or friendship or sexuality or authority or power, etc.
- Transcenders are interested in causes beyond their own skin and are better able to fuse work and play, and they are more interested in kinds of pay other than money. Mystics and transcenders have throughout history seemed spontaneously to prefer simplicity and to avoid luxury, privilege, honors, and possessions.

---

96. *The Farther Reaches of Human Nature.* Penguin Books (1971).

Maslow was recognizing a higher consciousness that went beyond the egoic consciousness of self-actualization. He was describing the shift from Stage 4 (self-actualization) to Stage 5 consciousness (self-transcendence). However, *self-transcendence* suggests that the self is no longer present; however, it is the *egoic self* that has been transcended, and a new kind of self that has emerged.

More than a decade before Maslow wrote about self-transcendence, Victor Frankl noted the limitations of self-actualization:

> ... the real aim of human existence cannot be found in what is called self-actualization. Human existence is essentially self-transcendence rather than self-actualization. Self-actualization is not a possible aim at all; for the simple reason that the more a [person] would strive for it, the more [they] would miss it. For only to the extent to which [people] commit [themselves] to the fulfillment of [their] life's meaning, to this extent [they] also actualize [themselves.] In other words, self-actualization cannot be attained if it is made an end in itself, but only as a side-effect of self-transcendence.[97]

Jim Collins, the researcher, author, teacher, and student of what makes great companies tick, introduced the concept of the *Level 5 leader* in his best-selling book, *Good to Great*:

> Level 5 leaders display a powerful mixture of personal humility and indomitable will. They're incredibly ambitious, but their ambition is first and foremost for the cause, for the organization and its purpose, not themselves. While Level 5 leaders can come in many personality packages, they are often self-effacing, quiet, reserved, and even shy.[98]

---

[97]. *Man's Search for Meaning*. Beacon Press (1959).
[98]. *Good to Great*. Harper Collins (2001). Collins' concept of the Level 5 leader predates by about 15 years the Macro-stages and Stage 5 I have proposed. I am merely pointing out the similarity.

After years of studying the highest-performing organizations, Collins found that all of them had Level 5 leaders in key positions. We now know this powerful finding applies widely across the spectrum of businesses, organizations, and social enterprise. It is a new paradigm of leadership that departs from the norms of the last 5,000 years. The Level 5 leader is not the egoic self that is still pervasive today, and they are nothing like the emperor archetype that has dominated throughout history. The Level 5 leader is a new and more evolved version of selfhood, with a deeper understanding of power and relationship, and we all have the possibility of moving in this direction. The evolution of the self is at the heart of the evolution of consciousness.

From the beginning of our lineage 3 million years ago, the brain has evolved within the social milieu. The nature of the self throughout our evolutionary history has been strongly influenced by the primary social unit that we lived in during each stage. As our ancestors evolved through the four primary stages of culture and consciousness, the size of social groups continually increased, in parallel with our co-evolving brain capacities. The expanding socio-cultural environment and our growing conscious capacities were inseparable and synergistic.

As the *Homo* lineage evolved, the primary social group grew in size from the *nuclear family* in the Mimetic stage, to the *clan* (also called the "troop" by some anthropologists) in the Magical stage, then to the *tribe* in the Mythical stage, and finally to the *nation* in our current Material stage. As we evolve further, the obvious next level of identity, beyond the nation, is the *planet*. When we truly grasp that we are embedded in a whole-planet system, our identity expands beyond the merely egoic, to become *planetary*.

It is unlikely that we can thrive or even survive on this planet unless we can see ourselves as a *species on a planet*, rather than nations pitted against other nations, groups against other groups, and people against other people. Nations can no longer stand alone, competing for territory and resources at the expense of others. Economies are globally intertwined; the oceans and atmosphere are shared by all of us; climate change and pandemics have no national boundaries. The Planetary consciousness is a higher awareness that allows us to transcend the separateness of nations, and the separateness of the egoic self, to experience a connected wholeness with all of life and Earth herself. We become *Earthlings* first and foremost, who happen to live in different regions we call nations, states, counties, cities, neighborhoods, and homes.

Planetary identity will be considered unpatriotic by people with a strong national identity. However, from the larger perspective, the planetary identity *transcends but includes* national identity in the same way that the United States, a nation, transcends but includes its individual states. These designations can still be helpful for managing infrastructures and sharing resources, but we don't need to have our personal identities anchored to any one of these, except the one at the top: the planet. Not long ago, people could not comprehend such a thing as *the planet*, but now with images from space and instantaneous global communications, we are waking up to the fundamental and essential reality that our home is a living planet.

The massive damage we are inflicting on the biosphere, and each other, stems from the Material-egoic consciousness that objectifies and separates everything in the world and seeks only selfish ends. But as Planetary consciousness emerges, national boundaries lose their importance, and the boundaries of the egoic self are softened as we begin to know and feel our *connectedness* with other people, with all of life, and with the planet.

5. Wholeness

Wholeness is a feature of the universe that physicist David Bohm discusses at length in his timeless work from 1980, *Wholeness and the Implicate Order*. The first chapter, titled "Fragmentation and Wholeness," lays out the case that human consciousness today is fragmented, and so is our view of the world. Fragmentation is at the heart of classical science and its reductionism that breaks everything into named categories and into parts that fit together in machine-like fashion. Even the human body can be thought of as a machine in the fragmented view of the world. But wholeness heals fragmentation by interrelating the fragments and erasing the boundaries. Bohm explains that how we see the world — whether fragmented or whole — determines our overall state of mind. Referring to humans in the masculine, he says:

> If he thinks of the totality as constituted of independent fragments, then that is how his mind will tend to operate, but if he can include everything coherently and harmoniously in an overall whole that is undivided, unbroken, and without a border (for every border is a division or break), then his mind will tend to move in a similar way.[99]

Gebser, speaking of the Integral consciousness, uses the term *diaphaneity* to mean "the transparent recognition of the whole, not just parts," and similarly states,

> The individual learns to see himself as a whole, as the interrelationship and interplay of magical unity, mythical complementarity, and mental conceptuality and purposefulness. Only as a whole man is man in a position to perceive the whole.[100]

Living organisms exemplify wholeness beautifully. Every cell and every organ grows and functions within the context of the whole organism. They cannot exist alone. A living organism cannot be taken apart and then reassembled according to instructions, as any machine can be. Our body is one whole entity that transcends and includes millions of intertwined subsystems that grow and resonate together.

We have already seen that fragmentation and separateness are at the very heart of the Material stage of culture and consciousness that solidified when the built world and the top-down power structures of civilization emerged fully 5,000 years ago. But humans are now capable of evolving

---

99. *Wholeness and the Implicate Order.* Routledge (1980).
100. The *Ever-Present Origin.* Ohio University Press (1985).

out of separateness and towards wholeness. Wholeness encompasses separateness, it transcends and includes separateness, so it is a higher dimension of consciousness. It is a marker of stage 5 Planetary consciousness that apprehends in *wholes* by integrating instead of separating. The planet is the ultimate whole.

Modern humans are thinkers. We live in our heads amidst an ongoing chatter, affectionately called the *monkey mind*, and the incessant inner voice that Carlos Castaneda called the *internal dialog*. We are overwhelmed by the fragmented excursions of our minds as our attention darts rapidly in short bursts and our thoughts race out of control. We can move this fragmentation towards wholeness by creating some space around all that thinking. This can be done by expanding our perspective and witnessing our own compulsive thinking (see *Meta-perspective*). We can create space around and within our thinking through awareness practices, such as meditation (see *Awakening*). This helps us get outside of our fragmented and frantic thinking and opens up the space behind and between thoughts, bringing fragmentation into wholeness.

An even deeper source of separateness, and lack of wholeness, is our troubled relationship with nature. The entire story of our lineage could be viewed as the process of separating from nature, as the *self* individuated and evolved, reaching a zenith in Material civilization. We must now move towards a relationship of wholeness with nature. This means seeing and feeling ourselves as part of the biosphere, as part of one whole system. This will not only be healing for *us*, but it will begin to heal the deep harm we have inflicted on the natural environment and other living things. For too long we have believed there are no consequences for our extractive industries, our enormous consumption of resources, and our wasteful lifestyles. We can no longer be *users* of nature and plunderers of Earth's natural treasures.

We must now move into a new role on this planet: to become the wise and benevolent caretakers of the planet and its living things. To do this we must be able to visualize the planet and all its inhabitants as an interconnected *whole*, a very big shift from centuries of damaging the biosphere.

The emerging Planetary consciousness sees the wholeness and connectedness and thus the *sacredness* of life and the planet.

Many people today live almost entirely in artificial environments made of concrete, plastic, and metal. We breathe polluted air and drink water that has been "treated" and recirculated many times; we eat food that is laced with artificial additives and stripped of nutrients. Many people are trapped in cities made of concrete and asphalt and have little or no idea that something called "nature" even exists outside this. We will re-establish a healthy relationship with nature when we realize and feel the interconnected wholeness of the biosphere. For inspiration and guidance, we can look to indigenous people and their sacred regard for nature. Ceremonies of appreciation and gratitude for the natural world awaken our connection with nature; spending time in nature, in beautiful natural places, awakens awe and reverence for the sacredness of our planet.

Before we can heal our relationship with nature and move towards wholeness, we must be brutally honest about our *footprint* on the environment — personally and as a society. We need to understand and be responsible for the real impacts of our excessive consumption of energy and unneeded products, and our dependence on a food system based on industrial farming and long-distance transportation.

It's easy to think that one person's impacts can't make much difference, but we always need to multiply by *a hundred million* or so to account for all the other people doing exactly what we are doing at the same time. Often our footprint is not immediately obvious — eating meat every day from industrial farms may not seem harmful until we look much deeper at all the impacts. An easy way for us to find out about this is to go to the *Global Footprint Network* (www.footprintnetwork.org) and calculate our own ecological footprint. Then we can decide what actions we are willing to take that will reduce our footprint on the biosphere. When we make commitments to small changes in our daily lives that reduce our footprint, it brings us closer to wholeness with nature.

As we look at all the people of the world, we see many differences among us — gender, age, skin color, ethnicity, political persuasion, religious belief, financial standing, education level, hair style, and much more. If we are in the mindset of fragmentation, we see how different we all are. But as we move into the wholeness of Planetary consciousness, we see how much we are the same. The differences are on the outside, like the clothes we wear. The sameness is on the inside, at the core of what it means to be *wholly* human.

6. Heart Opening

Traditional wisdom has long recognized the heart as the seat of love, courage, intuition, and truth, while Western science tells us it is merely a pump in the machine we call the human body. But extensive research now indicates that the heart is much more than a pump. It is a highly complex information processing center — a functional brain — that communicates with and influences the cranial brain via the nervous system, hormonal system, electromagnetic fields, and pressure waves in the circulatory system. In these four ways the heart affects brain function and all the body's organs, and it plays an important role in our mental and emotional experiences and in the quality of our lives.

Not long ago it was thought that neurotransmitters, like norepinephrine, epinephrine, dopamine, and oxytocin, were produced exclusively in the brain, but now we know that the heart also produces these in comparable amounts. The major neural connection between the heart and the brain is the *vagus nerve*, and most of its nerve fibers are *afferent*, meaning they carry impulses *toward* the brain. This means that the heart sends more information to the brain than the brain sends to the heart, suggesting that the heart "knows" many things *before* the brain does. Many of our first impressions of the world come from the heart, and later our neocortex receives that information and creates our conscious experience.

In neuroscience and in psychology, the emotional and the cognitive are thought of as separate but interacting systems that communicate via bi-directional neural connections between the neocortex and emotional centers, such as the amygdala. However, the neural connections that transmit information *from* the emotional centers *to* the cognitive centers in the brain are stronger and more numerous than those that convey information from the cognitive to the emotional centers. This is why emotions can so easily influence thought, but it is difficult to influence emotions by thinking.

The conventional wisdom of the last 400 years, arising from the Mental consciousness, is that emotions interfere with logic and rationality by "clouding one's thinking." But emotions have their own type of intelligence and have been shown to be critical in decision-making. The concept of *emotional intelligence*, popularized in the mid-1990s by Daniel Goleman,[101] adds an important dimension to the narrow view that intelligence comes only from the mental or intellectual domain. Qualities such as motivation, altruism, compassion, and self-awareness are found to be more important than a high IQ for navigating life's challenges, as well as the ability to self-regulate, to control impulses, and to self-direct emotions.

Decades of research at HeartMath Institute (www.heartmath.org) indicates that the key to the successful integration of mind and emotions lies in increasing one's emotional self-awareness. This supports the harmonious function, or *coherence*, among the neural systems that underlie cognitive and emotional experience. Heart rate variability (HRV) is one of the best indicators of coherence and overall health. We might think that our heart rate would be fairly constant when we sit quietly for a few minutes, but it actually varies naturally because of the complex interplay of incoming and outgoing signals from all over the body. When people are in physical, psychological, or emotional distress, HRV is irregular and erratic, but when people are experiencing positive emotions, such as appreciation, love, or caring, HRV is smooth and wave-like (6-2).

---

101. See, for example, Daniel Goleman's *Emotional Intelligence: Why It Can Matter More Than IQ*, Bantam Books (1995).

6-2

Courtesy of HeartMath Institute

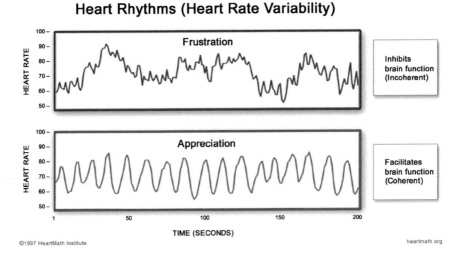

HRV is among the physiological indicators of a person's state of health and well-being that can be measured with sensors. Other vitals that can easily be measured include respiration rate, skin resistance, blood pressure rhythms, and multiple brain rhythms that show up on electroencephalograms (EEGs). Decades of experiments show that the heart has the ability to pull these other biological oscillators into synchronization, or *entrainment,* with its own rhythms. Researchers at HeartMath Institute conclude that:

> Positive emotions in general, including self-induced positive emotions, shift the entire system into a more globally coherent and harmonious physiological mode, one that is associated with improved system performance, ability to self-regulate, and overall well-being. When the brain and heart are out of sync, our nervous system signals are chaotic and we tend to get frustrated, anxious or angry more easily. It's like driving a car with one foot on the gas and the other on the brake. It's a jerky ride, we waste gas, and the car wears out faster.[102]

---

102. *Science of the Heart, Volume 2,* p 26. HeartMath Institute (2015).

Here is a simple exercise called Quick Coherence*, developed by HeartMath Institute, designed to increase heart coherence by generating positive emotions. It can be done for a few minutes every day or when stress and anxiety are high. When you are in a coherent state, your thoughts and emotions are balanced, and you experience ease and inner harmony.

> ### QUICK COHERENCE® TECHNIQUE *
>
> **Step 1: Heart-Focused Breathing**
>
> Focus your attention in the area of the heart. Imagine your breath is flowing in and out through your heart or chest area. Breathe a little slower and deeper than usual.
>
> **Step 2: Activate a Positive Feeling**
>
> Make a sincere attempt to experience a regenerative feeling, such as appreciation or care for someone or something in your life. Find an easy rhythm that's comfortable.
>
> * *Quick Coherence is a registered trademark of Quantum Intech, Inc. (dba HeartMath Inc.)*
> *https://www.heartmath.org/resources/heartmath-tools/quick-coherence-technique-for-adults/*

Heart coherence also extends outside the body and affects other people and living things. One way this happens is by means of the strong electromagnetic fields produced by the heart. The heart is the most powerful source of electromagnetic energy in the human body, producing the largest rhythmic electromagnetic field of any of the body's organs. This field, measured and displayed in the electrocardiogram (ECG), can be detected anywhere on the surface of the body. Furthermore, the magnetic field produced by the heart is more than 100 times stronger than the field generated by the brain and can be detected up to 3 feet away from the body, in all directions, using sensitive magnetometers.

Many experiments have demonstrated the coherence that can occur between a mother and a child, between two different people, and between people and animals. The graph in 6-3 shows the coherence between Ellen and her horse, Tonopah, after Ellen begins a "Heart Lock-In" exercise, which is similar to *Quick Coherence* described above.

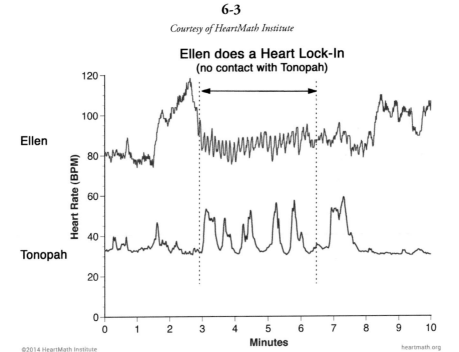

Courtesy of HeartMath Institute

Coherence, like that between Ellen and her horse, occurs in many human affairs. Anyone who has watched a championship sports team or experienced an exceptional concert knows something special is happening that transcends normal performance. It seems as though the players are in sync and communicating on an unseen energetic level. A growing body of evidence suggests that an energetic field can form between individuals in groups through which communication among all the group members occurs simultaneously. There is a coherent group field connecting all the members.

A subtle, yet influential electromagnetic or "energetic" communication system operates just below our conscious level of awareness. The nervous system and heart act as antennae tuned to the electromagnetic fields produced by other individuals, while our own heart produces electromagnetic emanations that others pick up. When someone is in the coherent state, they are more sensitive to receiving information contained in the fields generated by others, as well as radiating coherence.

Intuition, that mysterious kind of knowing without rational or deliberate thought, also comes from the heart. It plays an important role in business entrepreneurship, learning, medical diagnosis, problem-solving, healing, spiritual growth, and overall well-being. Intuition is *non-local* because the knowledge it brings us cannot be explained by past or forgotten knowledge or by sensing environmental signals. Non-locality transcends time and space as we know them in our normal head-centered Mental consciousness. The capacity to receive and process non-local information appears to be a property of all physical and biological systems because of the inherent interconnectedness of the unified field underlying material reality. (See *Connectedness*.)

Traditional wisdom holds that intuition comes to us through the heart. This may be because the heart is more closely coupled to the non-material field of information — the Akashic field — that is not bound by the classical limits of time and space. When our heart is more open, we know things directly before we consciously know them in our brain. A number of experiments have now been done[103] demonstrating that the heart receives and processes information about a future event *before* the event actually registers in the brain.

Heart intelligence is the synchronistic alignment, or coherence, of the heart with the cognitive mind and emotions, giving us greater access to our intuitive intelligence. The research at HeartMath Institute[104] suggests that intuitive intelligence is accessed more effectively by first getting into a coherent state, quieting mental chatter and emotional unrest, and paying attention to shifts in our feelings, a process that brings intuitive signals to conscious awareness.

---

103. *Feeling the future: A meta-analysis of 90 experiments on the anomalous anticipation of random future events*, Daryl Bem, Cornell University, et al. NIH online, https://www.ncbi.nlm.nih.gov/pmc/articles/PMC4706048/.
104. For details on these studies and all other research at HeartMath Institute, download the free e-book, *The Science of the Heart*, at https://www.heartmath.org/research/science-of-the-heart/.

We should think of the heart as both a parallel brain that interfaces with our cranial brain and a sense organ — our sixth sense. Our well-known five senses detect light, sound, chemicals, and so on, feeding into the cranial brain, which then filters and constructs the Material world, the explicate order of Stage 4 consciousness. But the heart is more attuned to the implicate order, and it lets us experience glimpses of higher dimensionality and non-locality, as a portal of sorts. Through the heart we can access the fields of information that tell us what's right and what's true, without the brain thinking about it.

We can arrive at some truths through the logic and rationality of the brain and the Mental consciousness. For example, if A = B and B = C, then it must be logically true that A = C. But we can also have a "feel" for what's true if we listen to our heart. Our head, filled with thoughts and rationalizations, can easily override what our heart is telling us, and wishful thinking often displaces what we know is true and right. Having a commitment to truth — to seeking it and facing it even when it's uncomfortable — is a core aspect of higher consciousness. Listening to and feeling our heart helps us know what is true and what is real.

Mainstream science is still deeply entrenched in the Material worldview that denies intuition because it can't be measured and explained within current paradigms. There are no equations for intuition. Yet there are many examples of scientists who made great discoveries by acting on intuition, including Einstein who famously said, "the intellect has little to do on the road to discovery. There comes a leap in consciousness, call it intuition or what you will, and the solution comes to you and you don't know how or why."[105]

The heart is an important part of our consciousness that is largely neglected and suppressed in today's Material/Mental world. But when we strengthen our heart awareness and open up our heart, we begin to connect with other hearts and also gain access to the non-material fields of the implicate order, where our mental faculties cannot go. By opening our heart and increasing

---

[105]. *Forbes Magazine.* September 15, 1974.

our heart intelligence, we can grow our intuition and experience more love, creativity, clarity, and truth. When we reclaim the heart as part of our consciousness, we feel first and then think, the ancient way of being human.

7. The Evolutionary Worldview

In Chapter One we explored the three great domains of evolution that modern science has investigated — the evolution of the universe, of life, and of culture — through the fields of astrophysics, evolutionary biology, and paleoanthropology. The cutting-edge knowledge of evolution in these fields of science was not available to our ancestors, or even our parents. We have called this *supra-evolution* to remind us that evolution is much more than just what happened to life on Earth, that it is woven into everything. Science has now recognized that the universe is not a predictable Newtonian machine made of material parts, but rather it is a creative evolutionary process of unfolding, an organism ceaselessly generating complexity and consciousness. This is the evolutionary worldview.

But even with an understanding of supra-evolution, we may not think that evolution is relevant in today's world, in our own lives, or for the future of humanity. There are two common assumptions that blind us to the importance of evolution for us today. The first is that evolution is very slow, taking millions and billions of years, so it plays no role in the lives of modern humans, or even in 5,000 years of human history. It happened long ago.

The second assumption is that humans have been fully modern and fully developed — a finished product — since the time of classical Greece. We often take the Greek tragedies written 2,500 years ago as evidence that "human nature" is largely unchanged because the characters are so much like us — they struggle with all the same issues and dilemmas that we do today. A corollary to this is that "history repeats itself" because human nature has stayed the same and we just keep repeating the same mistakes. From this view, war and violence will always be with us, and people will always be self-centered and tend to lie, cheat, steal, and be adulterous because this is human nature. We believe that we must resign ourselves to more of the same.

The evolutionary worldview pushes back on all this. As for evolution being slow, let's step back and notice, from the beginning, the trajectory and the pace of evolution. It took about 9 billion years of cosmic evolution for the universe to produce our planet (yes, that *is* slow); then about 3 billion years for life to evolve into animals. Then it took 600 million years for animals to evolve into apes, and only about 8 million years for apes to start making stone tools and expressing culture; then it took about 2.5 million years for our ancestors to master speech, and another 50,000 years for writing and civilization to emerge, then about 4,500 years for empirical science to be established, 300 years for quantum physics to be developed, and another 60 years for the world wide web to become ubiquitous. Clearly, the pace of evolution keeps accelerating. If we graphed this, it would look something like Figure 6-4.

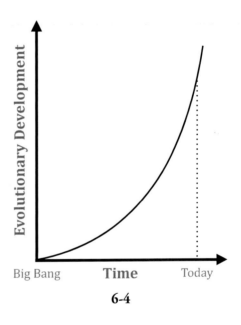

6-4

We should not confuse this accelerating pace of evolution with the exponential growth of technology and information that has occurred during our current stage of development. For example, Gordon Moore noticed in the 1970s that computing power was doubling about every two years, as costs went down with the miniaturization of transistors on computer chips. This was sanctified as "Moore's Law," and the trend has continued as he predicted until recently. But this is not due to evolution — it's a product of science, technology, and free markets.

The slope of the graph in Figure 6-4 is now so steep — the rate of evolutionary change is so high — that we certainly *should* expect evolution to be a palpable factor for us today. The evolutionary worldview embraces evolution as an active agent *right now*, for humanity and for each of us. There is no room in this view for "human nature" as a fixed thing, as something we're stuck with. History does indeed tell us that for the last 5,000 years humans have tended to be power-hungry, war-like, and jealous of others' possessions. But while the cynics will say, get used to it because that's how it will always be, the evolutionaries say, we can and we will evolve beyond those things. We are evolving into something new.

Evolution goes well beyond science when we bring in consciousness. As we have seen, science has struggled with consciousness because it cannot be measured, and we can't yet explain what it is or where it comes from. But all of us experience it intimately at every waking moment, and the founders of quantum physics were forced to confront it, so we should be able to talk about consciousness as something real. One purpose of this book is to suggest some basic language and structure for the scientific study of consciousness, and another is so that ordinary people can talk about it over a beverage of their choice.

We have reached the point in our evolution where we can *recognize* consciousness, and more so, we can now recognize the *evolution of consciousness*. At the very core of the evolutionary worldview is the understanding that consciousness is evolving — individually and collectively — and this is happening very rapidly today compared to the pace of evolution in the past. It is the evolution of consciousness that will carry humans forward and determine our future on this planet.

Within the evolutionary worldview we can now recognize four major stages in the evolution of human consciousness so far — the Macro-stages. These tell the story of the genus *Homo* in the broadest brushstrokes, as the makers of tools, the creators of culture and knowledge, the builders of civilization, and the rulers of Earth. But as magnificent as the rise of humanity may seem, the fact remains that we have now become a destructive force on this planet. We face a crisis of collapsing eco-systems and species extinctions, while our population continues to soar. If we blindly continue in the same way, degrading and assaulting Earth's biosphere, we are very likely headed toward catastrophic losses of human life from climate change, war, pandemic, and other scenarios we can't yet imagine.

But we have also evolved the ability to alter the course we are on and to shape our own future. The evolutionary worldview informs us that we are at the end of a major stage of history and development, and on the verge of entering a new regime of consciousness that sees the world in a completely different way. This is Stage 5 of human evolution, the Planetary consciousness, the next structure of consciousness that inclusively brings together all of the previous structures into a higher-functioning whole.

The critical question for us today is whether we can make the transition out of the Material regime of culture and consciousness, where the world is filled with lifeless objects that we can manipulate and exploit for our personal gain, and evolve into Planetary culture and consciousness, where the world is connected and interrelated in wholeness. The Material consciousness still overwhelmingly dominates human affairs on this planet today, so how can we be hopeful that this can change? Most people are just struggling to survive in the material world, so how can they concern themselves with higher consciousness?

Many people alive today will not be able to rise above the challenges of the material world by themselves. It will be up to a core group of pioneers to build a new culture supporting the evolution of consciousness in everyone — and it's already happening. This core group of evolutionaries is not exclusive — quite the opposite. It is open to anyone and everyone who understands the evolutionary trajectory of humans on Earth today, to everyone who is ready to move beyond self-centered materiality and into Planetary consciousness. If you are reading this book, you are likely among the pioneers of the emerging world.

The evolution of consciousness takes place in individual people, but when groups of people evolve, a shared consciousness emerges that manifests as culture. If Planetary culture and consciousness grow and spread to enough people, it will become the dominant paradigm. In the past this has taken thousands or tens of thousands of years, but today it could happen in a very short time.

We are at a fork in the evolutionary road, and it's ours to choose, or not choose, which way to go. One fork is the main road with business as usual: continuing to grow our population and to extract for profit as much as we can from Earth, trying to solve each new crisis with the mindset that created it, and putting band aids on grievous wounds. But this main road leads to disaster for humanity. We simply cannot go on destroying our planetary support systems.

The other fork is a faint path that many do not see. On this path we wake up to the planetary crisis and choose to influence our destiny by actively participating in our own evolution. There are many actions we can take to support and catalyze our evolution — individually and collectively — and this path will unfold if enough people can come together with a shared vision of a new world and shared commitments to right actions. We have the resources and the tools to follow this evolutionary path, but we must take action soon and in the right way. In Chapter Seven we will explore this work and inspire ourselves to become the agents of evolution who will help birth a new humanity and a new chapter in our story.

6-5

Seven Markers in a Nutshell

Awakening
- The "light goes on"
- The "wake up call"
- Bigger understanding
- Heightened awareness
- Conscious action: mindfulness
- Increase in knowledge
- Authentic learning

Connectedness
- Ubuntu
- Interbeing
- Oneness
- From mine to ours
- Ecology and relationship
- Resonance
- Unified Field
- Entanglement and non-locality
- Planetary consciousness

Meta-perspective
- Without a single perspective
- The view from above
- Witnessing one's own actions
- Changing perspective
- Being in someone else's shoes
- Recognizing conditioning
- Empathy and compassion
- Transcendence of paradox
- Loss of self-centeredness

Softening of the Ego
- Cessation of self-promotion
- From *me* to *we*
- Empowerment of others
- Collaboration and partnership
- Self-transcendence (Maslow)
- Level 5 leadership (Collins)
- Authentic self

Wholeness
- Integration of parts
- From machine to organism
- Systems and relationships
- Transcendence of categories
- Inclusion and commonality
- From differences to similarities

Heart Opening
- Heart-Brain coherence
- Knowing without thinking
- Heightened intuition
- Apprehension of truth
- Exchange of love
- Beauty and gratitude
- Heart connection with others

The Evolutionary Worldview
- Universe as process(es), not thing(s)
- Supra-evolution: cosmos, life, consciousness
- Macro-stages of human evolution
- Personal evolution of consciousness
- Planetary culture and consciousness

CHAPTER 7

Emerging World

Let us complete our journey together by envisioning a positive future for humans on this planet and exploring *what* we must do and *how* we can bring this about. This is the project of our time, to co-create a new civilization that serves all people and is good for our planet. It is in fact the most audacious and ambitious project humans have ever undertaken. And it is the most important. Sending humans to Mars pales by comparison. But the clock is ticking as we continue to grow our population, degrade our life support systems, and fight over vanishing resources. Things indeed look bleak today, but I have argued many times already in this book that humanity is now in the throes of a great transformation from one major historical stage to another, from one dominant structure of consciousness, the Material, to a new consciousness that allows us to see the world as a connected planetary whole.

We who are alive at this critical moment can take part in birthing a new civilization that operates within the planetary boundaries and produces well-being for all people and all of life — an *ecological civilization*. Humans are completely capable of achieving this. We have the knowledge, we have the technology, there are enough resources for everyone, and the only question is whether we have the wisdom and the will to take right action. What gives me even more hope and further fuels my optimism is that this vision of an ecological civilization and a world that works for everyone is strong in young people today, much more so than their elders. The new civilization is *their* world.

We begin this final chapter by exploring the requirements for an ecological civilization. What must it do, and not do, and what will it be like? In recent decades as our ecological crisis has unfolded, there have been widespread collaborative discussions and prolific writings about new civilization and new economics. Scholars and citizens from all over the world have awakened to the fact that our current civilizational paradigm cannot continue; civilization as we've known it will either destroy itself, or it will evolve into something new.

Of all that has been written and spoken about creating an ecological civilization, we will explore several efforts that I have found inspiring and representative of a large international consensus. I have chosen the work of two scholars, *Kate Raworth* and *David Korten*; the initiatives of two organizations, the *Club of Rome* and the *Well-being Economy Alliance*; and the blueprint set out in one document, the *Earth Charter*. As you will see, we have brilliant thinking and clear direction about *what* we need to do to co-create a sustainable and regenerative civilization. That should comfort us.

But then comes the *hard* question: *How* will we do it? How will we actually transform our Material civilization on a planetary scale into one that serves all of us and honors our sacred home, our planet? Where are the levers of change that can redirect the massive global power structures that still rule the world? I have concluded that our best and probably only hope lies with culture and consciousness and, of course, evolution. If enough humans can begin evolving into the Planetary consciousness, our institutions, our leaders, our businesses, our economy — everything — will follow. Success will depend on large numbers of people on Earth coming together around a shared vision of our future and making shared commitments to action. This has already begun.

20th Century Economics

Economists have been among the most engaged scholars in thinking about and designing a new civilization. Why economists? Because economies — at local, national, and global scales — determine nearly everything about what is possible for the people who live within our civilization and what our relationship with nature will be. Economic systems are human-made — they are sets of agreements and rules within a society about how the exchange of money, goods, and services will work and how resources will flow. Economic systems are a complex and dynamic interplay of many elements, including laws and governments, businesses and markets, producers and consumers, value and money, and, not the least, human psychology. All of us are enmeshed in economic systems. They are the fabric of civilization.

However, today's new generation of economists has realized that conventional economic theory as taught at most universities has become obsolete. It is mechanistic and based on flawed assumptions about supply and demand, human nature, the planet, and perhaps most important, the purpose of the economy. The modern economy of the West, which dominates the world today, is often said to have been launched in April of 1947 at Mont Pelerin, Switzerland where thirty-eight men and one woman met with the aim of designing the post-war world. These were mostly economists, including some who would go on to win Nobel Prizes and advise world leaders. After the first meeting, the group continued to meet regularly and came to call themselves the Mont Pelerin Society. Among their stated aims was "the creation of an international order conducive to the safeguarding of peace and liberty and permitting the establishment of harmonious international economic relations," and "re-establishing the rule of law and assuring its development in such manner that individuals and groups are not in a position to encroach upon the freedom of others and private rights are not allowed to become a basis of predatory power."[106]

That first meeting occurred just after World War II when fears of fascism and communism drove a unitary desire to promote and protect free societies and free markets. Although these original intentions were laudable, the actual economic realities that played out over the following decades were quite different. The post-war economy moved steadily away from serving *people* and increasingly towards serving big corporations and the wealthy. Those who promoted this trend asserted that when big corporations and the wealthy do well, it benefits everyone. This has come to

---

106. https://www.montpelerin.org/statement-of-aims/.

be called "trickle-down economics," and it was successfully implemented by Ronald Reagan and Margaret Thatcher in the 1980s. But it has now been demonstrated[107] that wealth has not trickled down to help everyone. Instead, this economic philosophy, still prevalent today, has insured that wealth will flow upwards to the already wealthy, while the assets and earning power of the bottom 99% shrink. Many young people today finish college will massive debt and see little chance of home ownership. Not much is trickling down for them.

A further problem with the twentieth-century economic philosophy still in use today is that it does not place value on the natural world, except for the resources that can be extracted for profit, and it relies heavily on growth and expansion. This combination has pushed us to the ecological limits of our planet and can only lead to disaster if we continue in this way. It is clearly time to put away the twentieth-century economic model and replace it with a new one for the twenty-first century.

Doughnut Economics

Kate Raworth calls herself a renegade economist, which means she rejects conventional economic theory and seeks new ways to address the social and ecological challenges we now face. Although formally trained in economics at Oxford University she also spent twenty years working with micro-entrepreneurs in the villages of Zanzibar, co-authoring flagship reports for the United Nations Development Program, and as a Senior Researcher at Oxfam. She is now a Senior Research Associate at Oxford University's Environmental Change Institute, Professor of Practice at Amsterdam University of Applied Sciences, and is a member of the Club of Rome. Her 2017 book, *Doughnut Economics: Seven Ways to Think Like a 21st Century Economist,* is internationally acclaimed, and her media work includes articles and interviews for *The Guardian, Financial Times, The Wallstreet Journal, Newsweek, The New Statesman,* CNN, the BBC, CBC, Al-Jazeera, and NPR. Go to **www.kateraworth.com** to find out more.

---

107. See for example: Dabla-Norris, Era, et al. *Causes and Consequences of Income Inequality: A Global Perspective.* International Monetary Fund (June 15, 2015).

Twentieth century economics uses Gross Domestic Product (GDP) as the primary indicator of economic health. GDP is defined as the total market value of the goods and services produced by a nation each year. However, GDP does not take into account the well-being and happiness of people or the health of our ecosystems and planet. Yet that's all that really matters for most of us. Kate Raworth realized this early in her studies of economics and began to rethink economics in 2011 with a sketch that became the visual frame of her now-famous *Doughnut Economics*. It consists of a pair of concentric rings (7-1), with the "safe and just space for humanity" between these rings, inside the doughnut. The inner ring, the *social foundation*, consists of basic human needs, such as food, water, sanitation, equity, and justice. Below (inside) this ring lies critical human deprivation and suffering. The outer ring, the *ecological ceiling*, represents the planetary boundaries, beyond which lies critical degradation of our planetary systems and possible tipping points. The doughnut serves as the compass for an economics that serves people and planet, rather than GDP.

7-1

THE BASIC DOUGHNUT

Courtesy: Kate Raworth

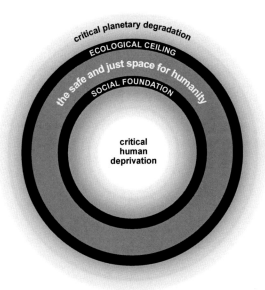

The *ecological ceiling* is made up of the following nine critical factors for planetary health that were identified in 2009 by an international group of Earth-system scientists led by Johan Rockström and Will Steffen: [108]

1. Climate Change
2. Ocean Acidification
3. Chemical Pollution
4. Nitrogen and Phosphorus Loading (from agricultural fertilizers)
5. Freshwater Withdrawals
6. Land Conversion
7. Biodiversity Loss
8. Air Pollution
9. Ozone Layer Depletion

The *social foundation* is made of the following twelve basic human needs specified in the United Nations' 2015 Sustainable Development Goals:

1. Food
2. Health
3. Education
4. Income and Work
5. Water and Sanitation
6. Energy
7. Networks (includes both social support and internet access)
8. Housing

---

108. Rockström, Johan, Steffen, Will, et al. *A Safe Operating Space for Humanity*. Nature, Volume 461, September 24, 2009.

CHAPTER 7: EMERGING WORLD

9. Gender Equality
10. Social Equity
11. Political Voice
12. Peace and Justice

Each of the nine elements of the ecological ceiling and the twelve elements of the social foundation can be measured (some more easily than others) to see if they lie *within* the doughnut's safe space for humanity or *outside* it — either beyond the planetary boundaries or below the thresholds of basic human needs. Raworth and others have gathered all this data for the entire Earth and put it into the doughnut, resulting in this:

7-2
EARTH'S DOUGHNUT OF SOCIAL AND PLANETARY BOUNDARIES AS OF 2017

Note: No global standards have been set for air pollution or chemical pollution.

Courtesy: Kate Raworth

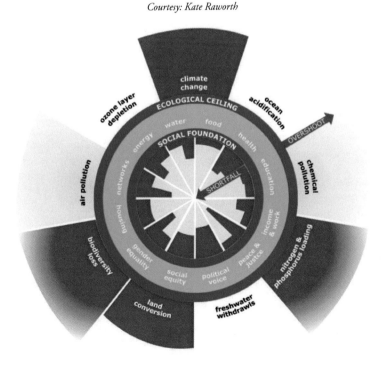

All the data behind this amazing graphic can be found in an interactive version at www.kateraworth.com/doughnut/, but here is a glimpse of the highlights. The four obvious areas in which humanity exceeds the planetary boundaries, shown in red *outside* the doughnut, are:

- **Climate Change.** Measured by parts per million (ppm) of atmospheric CO_2.

  Planetary Boundary: 350 ppm

  Current Value: 400 ppm and rising

- **Biodiversity Loss.** Indicated by species extinction rate, measured in species lost per million species per year.

  Planetary Boundary: At most 10

  Current Value: About 100 to 1,000 and rising

- **Land Conversion.** Measured by percent of forested land remaining since large-scale human deforestation began.

  Planetary Boundary: At least 75%

  Current Value: 62% and falling (worsening)

- **Nitrogen and Phosphorus Loading.** Measured in millions of tons per year of fertilizer applied to land.

  Planetary Boundary: Nitrogen - 62, Phosphorus - 6.2

  Current Value: Nitrogen - 150 and rising, Phosphorus - 14 and rising.

- **Note:** There are no global standards yet in place for air pollution or chemical pollution.

Some of the most glaring shortfalls in humanity's social foundation, shown in red *inside* the doughnut, are:

- **Peace and Justice** - 85% of people on Earth live in countries with excessive corruption.

- **Gender Equality** - The worldwide earnings gap between men and women is 23%.

- **Health** - 46% of people live in countries where the under-five mortality rate exceeds 25 per 1000 live births. 39% of people live in countries where life expectancy is less than 70 years.

- **Sanitation** - 32% of people do not have access to improved sanitation.

- **Networks** - 24% of people do not have someone they can count on for help in times of trouble. 57% of people have no internet access.

- **Energy** - 17% lack electricity. 38% lack access to cooking facilities.

- **Income** - 29% of people live on less than $3.10 per day.

Raworth's doughnut provides a clear assessment of our global civilization now, and it also shows us the way towards a new civilization in which all people thrive within the planetary boundaries. In her book, she uses the frame of the doughnut to set out "7 Ways of Thinking Like a 21st Century Economist," the book subtitle. These are seven guiding principles that are clean and simple and wise. Although I will not present all of them here (please read this book), I will mention Raworth's sixth principle: *Create to Regenerate*.

Designing regenerative systems means creating a *circular economy*. This concept was articulated over fifty years ago by economist Kenneth Boulding. Calling it the "closed economy" he stated with great prescience in 1966:

> I am tempted to call the open economy the 'cowboy economy,' the cowboy being symbolic of the illimitable plains and also associated with reckless, exploitative, romantic, and violent behavior, which is characteristic of open societies. The closed economy of the future might similarly be called the 'spaceman' economy, in which the earth has become a single spaceship, without unlimited reservoirs of anything, either for extraction or for pollution, and in which, therefore, man must find his place in a cyclical ecological system which is capable of continuous reproduction of material form even though it cannot escape having inputs of energy.[109]

---

109. Boulding, Kenneth E. *The Economics of the Coming Spaceship Earth.* Appears in Jarrett, H. (ed.), *Environmental Quality in a Growing Economy. Resources for the Future.* Johns Hopkins University Press (March 8, 1966).

Since the 1960s, Boulding's "closed economy" has become known as the circular economy and has been advocated by many leading economists and futurists, including Buckminster Fuller, Michael Braungart and Bill McDonough, John Lyle, Barry Commoner, Walter Stahel and Genevieve Reday, Tim Jackson, Robert Costanza, Paul Ormerod, Janine Benyus, Katja Hansen, Gunter Pauli, Amory Lovins, Paul Hawken, Hunter Lovins, Johan Rockström, and Kate Raworth.

Raworth also calls this the *butterfly economy*[110] because it can be depicted as shown below, right. Our current economics and lifestyles are linear (left) in that we extract and consume resources, then throw away large amounts of stuff in a one way drain of resources and accumulation of wastes. A circular economy reuses, repairs, and recycles, it recaptures and restores, thus regenerating, rather than depleting, our planet.

7-3

THE *LINEAR* ECONOMY IN CONTRAST TO THE *CIRCULAR* ECONOMY

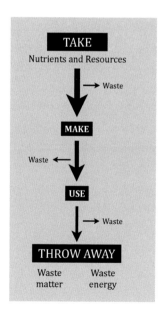
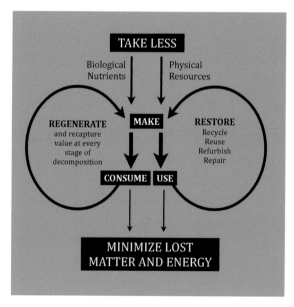

---

110. Introduced in 1998 by Paul Ormerod in his book *Butterfly Economics*.

Circular economics is just one example of new approaches we can take that will allow our civilization to operate within the ecological boundaries of our planet. These are not difficult or costly and will not depress prosperity. In fact, just the opposite. Many businesses have now found that circular business practices that reduce their ecological footprint actually lead to higher profits. There is indeed hope for an ecological civilization if we can just break out of old habits and old thinking. Let us take Kate Raworth's concluding remarks from *Doughnut Economics* to heart:

> Doughnut Economics sets out an optimistic vision of humanity's common future: a global economy that creates a thriving balance thanks to its distributive and regenerative design. Such an aspiration may seem foolish, even naïve, given the intertwined crises of climate change, violent conflict, forced migration, widening inequalities, rising xenophobia, and endemic financial instability we face. Watch or read the daily news, and the possibility of breakdown — social, ecological, economic, and political — feels very real. Humanity's glass easily looks half empty. Follow those fears through and you can quickly find yourself turning to the economics of collapse and survival which, like all powerful frames, could help to make those outcomes self-fulfilling.
>
> But there are enough people who see the alternatives, the glass-half-full future, and are intent on bringing it about. I count myself amongst them. Ours is the first generation to deeply understand the damage we have been doing to out planetary household, and probably the last generation with the chance to do something transformative about it. And we know full well, as an international community, that we have the technology, know-how, and financial means to end extreme poverty in all its forms should we collectively choose to make that happen.

The Ecological Civilization

David Korten is a former professor at the Harvard Business School, a political activist and critic of corporate globalization, a member of the Club of Rome, a prolific writer and author of many influential books,[111] co-founder of *YES! Media*, and, he says, "by training and inclination a student of psychology and behavioral systems." His work builds on lessons from the 21 years he and his wife, Fran, lived and worked in Africa, Asia, and Latin America on a quest to end global poverty. Korten's writings reflect a synthesis of today's best thinking on new civilization and new economics, the result of his decades of collaboration with scholars and leaders from around the world. He has kindly given me permission to use his words, which you can find more of at **www.DavidKorten.org**.

David Korten sums up our predicament today and our fundamental choice:

> We humans now consume at a rate 1.7 times what Earth can sustain and the richest 26 of us own wealth equal to that of the poorest 3.7 billion. We face a choice. We can continue the Imperial Era's drive to wealth concentration and environmental destruction. Or we can create an Ecological Civilization that secures material sufficiency and spiritual abundance for all, in balance with the regenerative systems of a living Earth.

He summarizes the work ahead in *Three Foundational Truths for a Twenty First Century Economics*:[112]

> A viable human future depends on the emergence of a new civilization guided by a twenty-first century economics. There are three basic truths, ignored or denied by our contemporary economics, that must serve as a foundation for a new economics.
>
> 1. **Humans depend on a living Earth.** All living beings depend on living communities that self-organize to create and maintain the conditions essential to all. Our well-being de-

---

111. See Appendix I.
112. From personal correspondence.

pends on the well-being of the living Earth that birthed and nurtures us. She long existed without us. We cannot exist without her. Restoring her health must be a defining priority of a twenty-first century economics.

2. **Humans are a choice-making species of many possibilities.** The historic diversity of human cultures and institutions demonstrates that we are a species of many possibilities. We can, for example, cooperate to nurture. Or compete to exploit. What defines our distinctive nature is our ability to make shared cultural and institutional choices that in turn shape our individual and collective choices.

3. **The drive to grow money imperils the human future.** Money is a number that has value only when accepted by people selling something we need or desire. Useful as a tool, money becomes dangerous when embraced as a purpose. A society that chooses to exploit people and nature to grow money for people who already have more money than they need grows the gap between rich and poor, destroys Earth's capacity to support life, and leads ultimately to human self-extinction.

The human future depends on making cultural and institutional choices that align with our needs as living beings, actualize the potential of our human nature, and make life, not money, our defining value. That will require the guidance of a twenty-first century economics grounded in these foundational truths and dedicated to the well-being of Earth and all its people.

Korten proposes the following *Eight Guiding Principles for the 21st Century Economics* that underlie the ecological civilization:[113]

---

113. From https://davidkorten.org/a-21st-century-economics-for-the-people-of-a-living-earth-version-4/.

Principle #1: The ***Purpose*** of the economy is to promote the *well-being of people and planet*, not the growth of GDP. If our priority is to meet the essential needs for food, water, shelter, and other basics for all the world's people, then we must measure for those results, and not GDP, so that we can get the outcomes we really want.

Principle #2: ***Resources.*** Use available human resources (labor) and natural resources only for purposes that enhance the well-being of people and Earth; eliminate uses that do not. Under this principle, war would be eliminated, as well as deceptive advertising, shoddy products, planned obsolescence, most agricultural chemicals, and the many other harmful things we do.

Principle #3: ***Labor.*** Direct the rewards and assign the rights of ownership to those who provide beneficial labor that produces community well-being. Transition to worker or community ownership combined with an ethical frame that recognizes our individual and collective responsibility for the well-being of the whole.

Principle #4: ***Money.*** Make the creation of society's money supply a transparent, publicly accountable process that serves the common good; not a secret process devoted to generating unearned profits for private bankers and financiers. In a modern society, those who control the creation and allocation of money control the lives of everyone. It defies reason to assume that society benefits from giving this power to global for-profit banks dedicated to maximizing profits for the already richest among us.

Principle #5: ***Education.*** Organize and manage education to support lifelong individual and social learning in service to the well-being of people and planet; end the isolation of the classroom and the fragmentation of disciplines. We must prepare youth for future leadership that builds on a moral foundation that recognizes our responsibility for one another and Earth, favors cooperation over competition, and prioritizes life over money.

Principle #6: ***Technology***. Create and apply technology only to enhance the well-being of people and planet; not to exploit, control, or displace people or nature for profit. Technology must be life's servant, not life's master. Humans have the right and the means to assure that technology is used only to serve humanity as a whole, not simply for the financial gain of a few.

Principle #7: ***Community***. Center economic life around self-organizing, self-reliant living communities and assure that all human institutions are accountable to and serve the well-being of the community in which they are located. Control of the economy must reside in inclusive, regenerative communities of people committed to working with one another and nature to grow and secure the well-being of all.

Principle #8: ***Population***. Seek a healthy balance among species, in support of the well-being of all; shed the conceit that Earth exists for humans. The health of any natural ecosystem depends on its ability to balance the populations of its varied species. We must now stabilize our population by maintaining free access to reproductive health care options and removing barriers to women in education and the workplace.

The Club of Rome

David Korten is not alone in proposing these principles for an ecological civilization. They reflect a wide consensus across the new civilization community and among his global collaborators. Korten is a member of the Club of Rome, a collection of notable scientists, economists, business leaders, and former politicians who began meeting in 1970 as the environmental crisis was first being recognized. Their stated goal "is to actively advocate for paradigm and systems shifts which will enable society to emerge from our current crises, by promoting a new way of being human, within a more resilient biosphere."[114]

---

114. https://clubofrome.org/about-us/.

According to the Club of Rome's website,

> In 1972, the Club's first major Report, *The Limits to Growth* was published. It sold millions of copies worldwide, creating media controversy and providing impetus for the global sustainability movement. This call for objective, scientific assessment of the impact of humanity's behavior and use of resources, still defines the Club of Rome today. While *Limits* had many messages, it fundamentally confronted the unchallenged paradigm of continuous material growth and the pursuit of endless economic expansion. Fifty years later, there is no doubt that the ecological footprint of humanity substantially exceeds its natural limits every year. The concerns of the Club of Rome have not lost their relevance.

David Korten describes the Club's current work:

> The Club of Rome was once a virtually all-male organization. In 2018 its members elected two women co-presidents. They immediately set about to strengthen and clarify the organization's priorities and restore its once significant contribution to the global dialogue on humanity's defining challenges.
>
> The Club is now focused on two primary initiatives. The first is a *Planetary Emergency Initiative* led by Co-president Sandrine Dixon-Decleve, whose significant connections and influence with European Union leadership could help achieve the carbon emissions reduction essential to human survival. The second is an *Emerging New Civilization Initiative* led by Co-president Mamphela Ramphele, revered leader of South Africa's anti-apartheid movement and co-founder of *Reimagine South Africa*. This initiative seeks a future that brings humans into co-productive balance with the regenerative systems of a living Earth while securing material sufficiency and creative opportunity for all. Rapidly unfolding environmental and social collapse provides a dramatic demonstration of the existential failure of existing economic institutions and theory. Yet decisive climate action cannot wait for the new economics or for the

systemic transformation of culture, institutions, technology, and infrastructure that such a civilizational transformation will require.

The *Intergovernmental Panel on Climate Change* has warned that to keep the global temperature increase below the target of 2.7 degrees Fahrenheit, we must reduce carbon emissions by 45 percent by 2030 and 100 percent by 2050. Decisive action must begin immediately using existing policy tools implemented through existing institutions. The *Planetary Emergency Initiative* focuses on that task. The longer term need for civilizational transformation will require far deeper, more complex, and less understood system changes and ways of thinking, including a new economics. Though still embryonic, serious discussion of what the deeper transformation will require is now underway. That a great many people are joining in gives me hope for the human future.[115]

The Well-being Economy Alliance

Another important effort to advance an ecological civilization is the *Well-being Economy Alliance*, or WEAll for short.[116] This is a global collaboration of organizations, movements, and individuals working together to transform the economic system into one that delivers human and ecological well-being. It provides the connective tissue between the many leaders in the movement for a well-being economy. WEAll now has over one hundred member organizations from all over the world and a stable of leading changemakers, including Ayabonga Cawe (South Africa), Robert Costanza (Australia), Sarah Deas (Scotland), Hunter Lovins (U.S.), Kate Pickett (Great Britain), Kristin Vala Ragnarsdottir (Iceland), Kate Raworth (Great Britain), Michael Pierson (U.S.), and Stewart Wallace (Great Britain), to name a few.

---

115. From https://davidkorten.org/my-priority-for-the-defining-year-of-humanitys-defining-decade/.
116. https://well-beingeconomy.org/.

This is how WEAll describes today's economy and the economy of the future:

> The economy is currently configured in a way that does not account for nature, in a way that is almost blind to distribution of resources, in a way that does not reward the best attributes of people, and in a way that puts measures of progress such as short-term profit and GDP to the fore. These are structures that have been designed — and hence can be designed differently, with a different purpose: that of collective well-being.[117]

WEAll states five basic requirements for assuring the well-being of all people and the planet:

1. **Dignity**: Enough for everyone to live in comfort, safety, and happiness.
2. **Nature**: A restored and safe natural world for all life.
3. **Connection**: A sense of belonging and institutions that serve the common good.
4. **Fairness**: Justice in all its dimensions at the heart of economic systems, and the gap between the richest and poorest greatly reduced.
5. **Participation**: Citizens actively engaged in their communities and locally rooted economies.

Of course, these perspectives are virtually identical to Korten's and Raworth's perspectives, and not by accident. The new economy community is highly aligned and collaborative, which is good news for all of us.

The Earth Charter

There are many other organizations and individual scholars who have arrived at the same vision of a well-being economy and an ecological civiliza-

---

117. All quotes from WEAll are from https://well-beingeconomy.org/.

tion. But perhaps the greatest collaborative effort, and the best summation of what we must attain, is the *Earth Charter* (www.earthcharter.org). The Earth Charter began as a project of the United Nations in the early 1990s, but it has been taken forward by the Earth Charter Commission, an independent international entity. After a decade of work, it was finalized in 2000 and has now been endorsed by over 6,000 organizations, including many governments. The Preamble by itself aligns with many themes we have explored throughout this book:

> We stand at a critical moment in Earth's history, a time when humanity must choose its future. As the world becomes increasingly interdependent and fragile, the future at once holds great peril and great promise. To move forward we must recognize that in the midst of a magnificent diversity of cultures and life forms we are one human family and one Earth community with a common destiny. We must join together to bring forth a sustainable global society founded on respect for nature, universal human rights, economic justice, and a culture of peace. Towards this end, it is imperative that we, the peoples of Earth, declare our responsibility to one another, to the greater community of life, and to future generations.

The Earth Charter is structured around the four broad principles below, and the full document expands each one of these in detail. Appendix V contains the full text of the Earth Charter.

1. Respect and Care for the Community of Life
2. Ecological Integrity
3. Social and Economic Justice
4. Democracy, Nonviolence, and Peace

The Charter ends with this vision: *Let ours be a time remembered for the awakening of a new reverence for life, the firm resolve to achieve sustainability, the quickening of the struggle for justice and peace, and the joyful celebration of life.*

From *What* to *How*

Whether we look to the *Earth Charter*, or the work of Kate Raworth and David Korten, or the initiatives of the Club of Rome and the Well-being Economy Alliance, we see the same themes and the same vision for a new civilization. This is the foremost thinking on the planet right now, reflecting deep compassion for humanity, a sacred regard for nature, and the highest levels of research and scientific knowledge. It should be reassuring to us that there is such strong consensus and alignment. Clearly, we know *what* we must achieve; the blueprints are there. But *how* can we do it, and why hasn't it happened yet after sustained efforts over the last fifty years? There has been tremendous energy and intelligence and unity of purpose behind the movement to create a well-being economy and an ecological civilization, but something is stopping it. Is anyone *against* this vision? What stands in the way?

To answer this, let's return to the perspective of the Macro-stages, and in particular, the Material stage. The current economy that concentrates wealth and power while exploiting people and nature is a primary feature of the Material stage of culture and consciousness that humans have been operating in for the last 5,000 years. Who could be against a well-being economy that lifts everyone? With few exceptions, it is those *with* power and wealth who don't want to lose it. Civilizational power structures are designed to maintain themselves and to resist attempts at change. We are still firmly entrenched in the power-centric civilizational structures and mindsets of the *Material* stage — the Imperial Era, as Korten calls it — and this is why we do not have a well-being economy. And there is little hope of changing this by using the existing means of the Material paradigm. We cannot solve this problem from within the consciousness that created it, to paraphrase Einstein. The solution, and the hope for the future of humans, is that we can evolve beyond the Material consciousness that is behind virtually all of our institutions, infrastructures, and problems. The most effective strategy for creating the world we want is to focus our efforts on growing the Planetary consciousness.

How can we influence the mega-structures of power that now rule the world? It seems impossible to change behemoths like global oil cartels, the pharmaceutical industry, big tech, or a U.S. Federal government that serves corporations over people. Yet there are glimmers of hope in some places. New Zealand has implemented the world's first well-being budget, Scotland has passed climate change legislation with the world's most ambitious goals, and Iceland has developed a well-being economy framework. Cities like Amsterdam are adopting policies based on doughnut economics. The Scandinavian countries have all moved towards well-being economies and are arguably the highest functioning democracies with the highest quotients of well-being in the world today. What has made these things possible? Quite simply, the evolution of consciousness. In all these places Planetary consciousness is emerging among those in power and the people who elected them.

How can these fledgling, isolated efforts grow and spread quickly? According to the Well-being Economy Alliance, *what is vital is a critical mass of people and organizations coming together to form a new dynamic movement to influence and inspire. This way collective impact is substantially multiplied. WEAll's role is to help catalyze this multiplication process.*

WEAll has developed a change model, or strategic vision, that guides their work toward a well-being economy. Three key ingredients are identified for bringing about change: *knowledge, narrative,* and *power base*. To elaborate:

> Knowledge. Grow and synthesize the well-being economy theoretical base that is now disparate and relatively hard to access; assemble best practices and proactively disseminate this knowledge; demonstrate the effectiveness of well-being economy approaches on small and large scales.
>
> Narrative. Create and disseminate positive new narratives about how we want to live together. Tell stories about what is right and what we want, instead of what is wrong and what we are against.

Power Base: Develop collaborative, cross-sector networks that build on agreements around goals, values, and principles. Strengthen partnerships and cooperation among aligned entities and individuals.

The Culture Project

All three of these strategies for change — knowledge, narrative, and collaborative networks — are aspects of *culture*. Therefore, quite simply, the way we will transform our economy and our civilization will be through consciously and strategically creating a *new culture* on a global scale, a culture that values the well-being of all people and the planet above all else; let us call this *positive* culture, or *Planetary* culture. We now have the ability, and a golden opportunity, to do this by using the powerful and ubiquitous electronic technologies we currently have, riding on the internet. We can create culture more easily and more powerfully than ever before. Therefore, the project of creating an ecological civilization and a well-being economy must become the *culture project*.

Guiding our efforts to create Planetary culture must be *right purpose*. Learning right purpose is the primary lesson for humanity at our evolutionary stage and the single factor that will determine whether we will evolve further or stall out. Right purpose is what Arthur Young meant by learning from the causality and determinism of Stage 4 and applying that knowledge correctly. We have learned the laws of physics and chemistry, and the laws of economics and business, and the laws of psychology and persuasion, but can we apply them correctly? It is critically important that we learn the right purposes for our knowledge and our tools, such as nuclear technologies, genetic engineering, and digital media, which have become so powerful they can bring harm and death when used for wrong purposes.

We must be guided by right purpose in everything we do, but what *is* right purpose? Who is to say? Don't individuals have differing perspectives on what right purpose is? After all, Hitler must have thought he had right purpose. But we must cast off moral relativism and "create your own reality," and let our hearts tell us what is right and true. From that perspective I propose the following simple definition of *right purpose*:

Right purpose *maximizes well-being* and *minimizes harm* for people and the planet.

Here, *Planet* means the entire planetary system of all living things and the biosphere, the totality of Gaia herself. *Well-being* and *harm* are at opposite ends of a spectrum. In practice, evaluating well-being is not always black or white, either/or, but we can place any action or purpose on this spectrum. For example, our current economic system that flows wealth to the wealthy and devalues nature produces far more harm than well-being, so we can conclude that this economic system is not structured for right purpose.

The well-being of people and planet is already the mantra of the new economics community and others working to build an ecological civilization, and it's an easy phrase to carry with us and apply as we make our own decisions and choose our actions. Will our choices and actions bring *well-being*, or will they do *harm*? If we ask this question honestly, from our heart, we will find the right purpose for a business, or a non-profit, or a government, or a life.

We can create a Planetary culture that inspires and uplifts people, empowering each of us to find our own unique gifts that generate well-being in the world. One of the most effective ways to create cultural meaning is through the Mythical consciousness, which is highly active in all of us. Stories and narratives are still one of the most powerful ways to convey cultural memes (elements of meaning). We can broadcast new narratives globally using the state-of-the-art technologies available to us. We can send out compelling stories that instill positive values and teach about our connectedness with nature and other people. Our new cultural narratives can teach about harm and well-being and present heroes who model kindness, wisdom, and compassion. We can do this using film, art, music, books, television, streaming media, live performance, and whatever new comes along. The culture project, then, is the effort to generate positive and uplifting culture that will engulf humanity, giving people a better alternative to the harmful and disempowering memes and narratives that are broadcast today.

But the same old question lingers: HOW can we actually do these things, to create a Planetary culture that promotes well-being? I will offer some concrete strategies and actions at the end of the chapter, but first we must explore the landscape of the "cultural work" ahead. We begin by considering the harm that has resulted from our Material culture and the healing that must begin so that our culture can evolve away from doing harm and towards generating well-being.

Healing from Materialism

As we envision and build a Planetary culture, we must keep in mind the importance of healing. We must create cultures of healing. What is it we need to heal? We need to heal the harm and damage from 5,000 years of *man over nature* and *man over man*, from the destructive and dehumanizing power structures of the Material culture. Although the age of civilization brought spectacular advances in knowledge and technology, inspiring art and architecture, and magnificent accomplishments like going to the moon, it has also resulted in deep harm and profound suffering. Now, we must acknowledge this truthfully and fully recognize harmful practices that continue, and we must commit to ending them. Then healing can begin.

The well-documented history of the last 5,000 years is filled with conquest, rape, pillage, genocide, and torture. Throughout history many people and groups of people have suffered profoundly at the hands of dominant society and the many cruel and ruthless emperors, kings, and governments. Although you and I were not complicit in the atrocities of the past, our ancestors probably were, and many of us today have benefitted from conquest. People in the United States, for example, must look honestly at their history and relationship with Native Americans and African Americans and seek to heal the profound harms and injustices that have been inflicted on these people. We cannot change what happened, but we all share the responsibility to confront the facts and to tell new stories that are truthful. More so, we must squarely confront the injustice and harm that continues today. The murder of George Floyd at the hands of police, filmed by a bystander and posted to the internet, awakened the world to the profound harm still being inflicted every day, especially to people of color. We must commit to building a world where these injustices and atrocities cannot happen.

All nations have their own dark past, and it's time to shine light in all these places; our new culture and our new narratives must be based on acknowledging the truth. Nazi Germany, for example, was one of the most advanced societies in all of history when it methodically murdered seven million people who were deemed undesirable. We must never forget what happened and what this tells us about humans at our current stage of development. It would be easy to put all the blame for the Holocaust on Germans and think the rest of us are above such a thing, but in actuality it is nothing new, and it could happen again, anywhere, unless we can be truthful, and learn, and evolve.

To build a healing culture we must create new historical narratives based on the facts from our past. Let us consider, as one example, the European colonization of the world, between about 1500 and 1900, and the harm it brought to non-Europeans. This "Age of Discovery" was formally motivated and justified by the *Doctrine of Discovery*, a decree issued in 1493 by Pope Alexander VI. The decree justified the seizure of land not inhabited by Christians, and it promoted Christian superiority and the "civilizing" of indigenous people. This launched a relentless competition between the British, the French, the Spanish, and the Dutch to claim and establish as many colonies as possible throughout Africa, Asia, Australia, New Zealand, and the Americas. The extraction of resources from colonies fattened the treasuries of the monarchs.

Conventional history records the bare facts of the age of colonization, such as this standard notation on Hernán Cortés from Wikipedia that we might learn in history class:

> Hernán Cortés was a Spanish Conquistador who led an expedition that caused the fall of the Aztec Empire and brought large portions of what is now mainland Mexico under the rule of the King of Castile in the early 16th century.

But the reality on the ground for the Aztec people is described more truthfully and fully in this account of what Cortés and his men did at the siege of the Aztec capital, Tenochtitlan, in 1521:

> Several hundred thousand were killed in the campaign including warriors and civilians. As many as 40,000 Aztec bodies were floating in the canals or awaiting burial after the siege. Almost all of the Aztec nobility were dead, and the remaining survivors were mostly young women and very young children. At least 40,000 Aztec civilians were killed and captured.
>
> After the Fall of Tenochtitlan the remaining Aztec warriors and civilians fled the city as Cortés' men continued to attack even after the surrender, slaughtering thousands of the remaining civilians and looting the city. They did not spare women or children: they entered houses, stealing all precious things they found, raping and then killing women, stabbing children. One source claims 6,000 were massacred in the town of Ixtapalapa alone. In the decades following the surrender of the Aztecs, roughly 7 million Aztec people were killed — 85% of the population.[118]

By some estimates the Native population of the Americas declined by 50 million people, or about 90% percent, because of European contact. Much of this was due to diseases, like smallpox, carried by the European colonizers and sometimes spread intentionally by offering infected blankets as gifts. By the time the thirteen British colonies became the United States, the "Indians" were a major problem that stood in the way of westward expansion. In an 1823 Supreme Court decision, *Johnson v. McIntosh*, the Doctrine of Discovery from 1493 became part of U.S. Federal law, legalizing the taking of Native peoples' land. In a unanimous decision, Chief Justice John Marshall wrote "that the principle of discovery gave European nations an absolute right to New World lands and native peoples certain rights of occupancy."

---

118. Leon-Portilla, Miguel. *The Broken Spears: The Aztec Account of the Conquest of Mexico*. Beacon Press (1992). See also: Naimak, Norman. *Genocide: A World History*. Oxford University Press (2017).

The *Monroe Doctrine* built on this decision by asserting that the U.S. was destined to control all land between the Atlantic and Pacific and beyond, a view known as *Manifest Destiny*. This gave colonists full permission to slaughter and take away the land from millions of Native Americans. Those who survived were marched onto reservations — lands considered worthless. The children of reservation families were taken away forcibly and put into boarding schools where all attempts were made to erase their culture, a practice that continued well into the twentieth century.

The truth of history is buried when we entertain ourselves with fictionalized narratives like "cowboys and Indians," where the savages attack innocent settlers and then must be killed. We all must acknowledge the genocide of the First Nations peoples and the intentional obliteration of their cultures. Those of us of European descent are told the story of colonization as though it was some natural and inevitable privilege of white people. But if a man entered his neighbor's home and killed the father, raped and killed the wife, captured the children to be sold as slaves, then helped himself to anything of value in the home, we would be shocked and dismayed that someone could do such a thing. Yet this is what the European colonization of the world was like. This is what the truth looks like.

Native American writer Edgar Villanueva expresses the trauma and harm experienced by all people of color:

> Imagine that all your family and friends and community regularly experienced traumatic events: upheaval, violence, rape, brainwashing, homelessness, forced marches, criminalization, denigration, and death over hundreds of years. Imagine the trauma of this has been reinforced by government policies, economic systems, and social norms that have systematically denied your people access to safety, mobility, resources, food, education, dignity, and positive reflections of themselves. Repeated and ongoing violation, exploitation, and deprivation have a deep long-lasting traumatic im-

pact not just at the individual level — but on whole populations, nations, and tribes. [119]

As if this was not enough cultural work, people of the U.S. also have the legacy of slavery to heal. Today, most people are at least vaguely aware of the cruelty and injustice of the slave trade and the slave culture that powered the plantation economy of the South until the 1860s. But we've been told that the African American slaves were freed after the Civil War, so everything's fine. In reality, the laws and constitutional amendments of Reconstruction, intended to guarantee African Americans the full rights of citizenship, were completely subverted and as a result the Civil Rights movement was delayed by a century. The inhumane treatment and oppression of African Americans and all people of color has continued to this day.

According to statistics compiled by the Tuskegee Institute, between the years 1882 and 1951 some 3,437 African Americans were lynched in the United States. Federal legislation outlawing lynching was introduced many times in Congress starting in 1918, but it always failed to pass because of opposition by Southern senators. Finally, in February of 2020, the Emmett Till Antilynching Act passed the House of Representatives but at this writing still awaits approval by the Senate and President before it becomes law.[120] Emmett Louis Till was a 14-year-old African American boy who was kidnapped, tortured, and murdered in 1955 after he allegedly whistled at a white woman in Mississippi. The men who committed this act were tried and found *not guilty*. Fifty-two years later, in 2007, the white woman admitted she had fabricated her story. There are many other incidents like this in the history of the U.S. and the world.

A snapshot of life today for African Americans is captured in incarceration rates. Research published in 2015[121] revealed that:

---

119. *Decolonizing Wealth.* Barrett-Koehler (2018).
120. Still pending as of 2/1/21.
121. Tucker, Ronnie B., Sr. *The Color of Mass Incarceration.* Ethnic Studies Review, Vol 37-38, No.1 (2014-2015).

- From 1980 to 2008 the number of people incarcerated in America quadrupled from roughly 500,000 to 2.3 million people, giving the U.S. the highest incarceration rate in the world. African Americans now constitute 1 million, or 43%, of the total incarcerated population.

- The incarceration rate of African Americans is now six times that of whites.

- African Americans are sent to prison for drug offenses at ten times the rate of whites, even though rates of drug use are nearly identical.

- One in every eight African American men is behind bars today, one in a hundred for women.

- African Americans make up only 12% of the United States population, but, together with Latinos, comprise over 60% of total inmate population.

How do we heal such deep harm brought about by our legal system? Zach Norris, Executive Director of the Ella Baker Center in Oakland and author of *We Keep Us Safe,* articulates beautifully the vision of a legal system and communities based on *restorative justice*:

> Rather than asking: "What law was broken, who broke it, and how should they be punished?" restorative justice asks, "Who was harmed? What do they need? Whose responsibility is it to meet those needs?" Restorative justice places the needs of the survivor at the center. It avoids terms like "victim," "perpetrator," and "offender," because of the way in which these labels can stick to a person forever and deny them the ability to evolve, to heal, and to change. ... Relationships, accountability, and healing are the principles at the foundation of the movement for restorative justice. [It] is based on holding people accountable, because they are held in community. ... Sending people to remote prisons and into isolation is often described as "tough on crime." In reality, this is the opposite of hardcore accountability. Prisons and cages just remove you, cutting you off from the fabric of community. But to have

to stay, to be forced by your entire community to hear the pain, trauma, horror, fear and grief you caused, that is hardcore. Real accountability comes when a person has to face what they have done, when they have to own up to it, and work to make it right.[122]

Restorative justice is not a theory or a futuristic dream — it is being used successfully today in schools and communities around the world. Nor is it new. Indigenous communities from the Native American to the Māori of New Zealand have used these principles of healing and accountability for millennia.

Healing the harms from the past is not solely a problem in the U.S. — there are endless examples of injustice and harm to people around the world throughout history, all a part of the 5,000-year-old Material stage we need to put behind us. The South African *Truth and Reconciliation Commission* that began in 1995 sought to address and heal the harms from apartheid, and it now serves as a model for truth commissions around the world. We cannot move forward until we are willing to acknowledge the truth about our past and commit to a new vision for the future. We need new stories that tell the truth and seed the vision of a Planetary culture where the well-being of every person matters.

A healing culture must also repair the harm we have inflicted on the natural environment of our planet. In the Material culture we have objectified Earth's living systems as inert and available for us to exploit endlessly, without consequences. We dump millions of tons of plastics into the oceans every year, and now plastics can be found throughout the marine ecosystem and inside many marine animals. We discard over *50 million tons* every year of electronic waste, like computers, cell phones, and TVs. How long can we keep doing that? We have fouled the air, polluted the waters, paved our prairies and forests, and decimated habitats at an alarming rate. This is not sustainable.

---

[122]. From *We Keep Us Safe. Building Secure, Just, and Inclusive Communities.* Beacon Press (2020).

Human destruction of wild habitats has produced an extinction rate that is now estimated to be 100 to 1000 times the average rate of extinction over the last few million years. One study published in 2017[123] found that the total population of animals on Earth today is only 50 percent of what it once was; it concluded that a *mass extinction event* is now underway, the result of human activities. Many other recent studies have reached the conclusion that we are squarely in the midst of the *sixth* mass extinction event in the last 500 million years — the last one eliminated the dinosaurs 65 million years ago. But this is the only mass extinction event directly caused by one living species (that would be us).

Another study published in 2018 found that humans, comprising a tiny fraction of all life, have destroyed 83 percent of wild mammals since the beginning of civilization.[124] Today extinction threatens 40 percent of amphibians, 30 percent of reef-building corals, more than 30 percent of marine mammals, and 10 percent of all insects. Domesticated animals (mostly cattle and pigs) now make up 60 percent of the biomass of all mammals on Earth, followed by humans at 36 percent. That is to say that only 4% of mammals on Earth today are wild! These are simple but painful facts about the harm we are doing to our planet and the life it supports. Only if we face these facts can we change them.

The Material worldview objectifies everything — from people and other living things, to mountains and oceans — as though they are all inert and disconnected from us. The healing of the biosphere will come naturally when we recognize the interconnectedness we are all a part of and learn to participate harmoniously within Earth's living systems.

The good news is that natural environments *can* heal by themselves if we leave them alone, if we simply stop assaulting them. Let's stop worshipping growth and urban sprawl and set aside natural open spaces that are good for both nature and people; we must preserve as many of Earth's remaining

---

123. Ceballos, Gerardo; Ehrlich, Paul; Dirzo, Rodolfo. *Biological annihilation via the ongoing sixth mass extinction signaled by vertebrate population losses and declines.* Proceedings of the National Academy of Sciences (25 July 2017).
124. Baillie, Jonathan; Zhang, Ya-Ping. *Space for nature.* Science (14 Sept 2018).

wild places as possible. Forests and other plant communities are far more valuable left alive and intact as CO_2 sinks, oxygen sources, and habitats for animals, as well as healing places for stressed humans.

National parks, state and provincial parks, and other public lands are immeasurably valuable resources, but they're constantly under threat from urban development, mining, grazing, and other resource extraction industries. But beyond the usual blaming of corporations, each one of us bears responsibility because of our own patterns of consumption and spending. We create the markets that support harmful industries. The reason that almost 60% of all mammals on Earth are cows and pigs is that so many of us *choose* to buy and eat meat all the time.

Our personal choices matter. If each of us chose to eat *less* meat, for example, the markets would shrink and industrial animal farming would decline; we would also be reducing the massive pollution and resource use — water, energy, and plant protein — associated with the meat production industries. If we chose to preserve as much natural space as possible, all over the planet, we would be taking a big step towards healing the biosphere. If we all chose to look honestly at our own resource consumption and our own footprint on nature, if we decide to stop being wasteful and begin using only what we *need*, there can still be abundance for all people *and* well-being for the planet. These things *will* happen naturally as our Material culture shifts towards a Planetary culture and consciousness.

Planetary Spirituality

As humans evolve out of the Material paradigm of reality in which external objects are primary, we will evolve a new spirituality. But this will not be the spirituality of the Magical or Mythical cultures, or the institutionalized religions that arose with the Material consciousness and civilization. It will be a new spirituality that integrates ancient wisdom with modern science. We will call this *Planetary spirituality*.

But what does *spirituality* mean? By this I mean a personal, felt experience; not a teaching or a set of practices and rules; not a program you can go through, or a how-to book. It is the direct, non-intellectual experience of reality in moments of heightened aliveness, a 'peak experience' as Abraham Maslow called it, that involves both the body and the mind. Buddhists call this heightened awareness *mindfulness*, and they emphasize that it is deeply rooted in the body.

Spiritual experiences evoke *feelings*, like wonder and awe, humility and gratitude, love and beauty, and well-being. This is in sharp contrast to the self-centered Material experience that brings us restlessness, dissatisfaction, boredom, alienation, loneliness, unfulfilled desire, and suffering — all the things the Buddha was addressing 2,500 years ago in response to the burgeoning Material consciousness at that time.

A spiritual experience is often called *mystical* because it is an encounter with mystery, or that which is beyond rational explanation. For thousands of years spiritual teachers and mystics have described a profound and ineffable sense of belonging to the cosmos as a whole, of connectedness or *interpenetration*, as Zen master Daisetz Suzuki called it. But many of the greatest scientists have also been inspired by mystery, as Albert Einstein famously expressed:

> The fairest thing we can experience is the mysterious. It is the fundamental emotion which stands at the cradle of true art and true science… the mystery of the eternity of life, and the inkling of the marvelous structure of reality…[125]

We can have spiritual experiences in many contexts — while playing or listening to music, while viewing art or participating in a physical activity like running or golf. Our thinking stops, if only temporarily, and our awareness of time seems to disappear. Many performers speak of peak experiences in which everything goes right, without effort. This has been called the *flow state*, or *the zone*, but it is nothing more, or less, than a spiritual experience.

---

[125]. *The World as I See It*. Philosophical Library (1949).

Religion originates in spirituality, but in actual practice nearly all organized religions have become elaborate institutions seeking to perpetuate themselves above all else. In the first thousand years of Christianity, for example, the principle figures were mystics who asserted the ineffable nature of religious experience and expressed their interpretations in terms of symbols and metaphors. A scholastic period followed, as the Mental consciousness came to the forefront, in which theologians formulated the Christian experience intellectually in dogmatic language, requiring followers to accept their teachings as the literal truth. Christian theology became more and more rigid and fundamentalist and devoid of authentic spirituality.

Institutionalized religion is a top-down power structure typical of the Material culture, and it is clear that millions of people are attracted to dogmatic religion that writes the script for them. When life is hard, this may seem to be a comfort. But it also means we forfeit our greatest potential — to be an empowered agent of action in the world and to express our own unique gifts (see *The Planetary Human* below).

Instead of obeying external authorities, we must be unafraid to look within and find our own light, to listen to our own heart that tells us what is true and right. Spirituality, or that which transcends the Material, is *within*, not *out there*. For many people spirituality centers on a relationship with God, or the Great Spirit, or the Life Force, or Nature. It can take any form as long as it generates well-being.

Authentic spirituality also requires a commitment to truth — that truth *does* exist — and a desire to search for it and to face it. We must relentlessly ask any and all questions, yet this is the very thing most religions[126] forbid: do not ask *why*, just obey. This is a fundamental violation of our humanness.

Planetary spirituality has one core value all people can share: the *sacredness of life and the planet*. We've lost sight of this sacredness in the Material

---

[126]. Judaism has a long tradition of encouraging questioning, seeking truth, and valuing education. Some would argue that Judaism is not a religion, but rather a *culture* based on community and personal relationship with God. (Interview with Rabbi Michael Kosacoff.)

culture, and this blindness is at the heart of our ongoing destruction of the biosphere and our disregard for people who are suffering. Indigenous cultures have historically been centered around the sacredness of life and Earth because their existence depends on it. Civilized people are waking up to the reality that ours does too. Stuart Kauffman expresses this beautifully:

> ... The sense of the sacredness of all of life and the planet can help orient our lives beyond the consumerism and commodification [of] the industrialized world, ... and heal the split between reason and faith, heal the split between science and the humanities, heal the want of spirituality, heal the wound derived from the false reductionist belief that we live in a world of fact without values, and help us jointly build a global ethic. [127]

We must now recover this ancient wisdom and recognize that the entire planet is one living entity, *Gaia*. Our planet is more than a just spinning ball in space, and it is not just a "rocky planet" as astronomers classify Earth. It is a massively interconnected living system of systems. From the tiniest microbes, to the teeming masses of humans alive today, to the entire biosphere and the dynamic geologic processes that build mountains and move continents, we are part of one whole living entity. Planetary spirituality is the realization of Gaia, our living planet, in which we are embedded. Science tells us this is true, and so does traditional wisdom.

In Material consciousness, we know only the explicate order of lifeless objects in the world around us, and we believe objects (including money) lead to happiness and well-being. When the sacredness of life and Gaia becomes our core value, as we move into Planetary spirituality, it will be impossible to have war, or industrial animal farms, or big game trophy hunting, or toxic industries that pollute the air and water.

Planetary spirituality universally recognizes the *connectedness* of all life. This has always been the core worldview of indigenous people and a fundamental tenet of many Eastern traditions. Now modern science has sub-

---

127. From *Reinventing the Sacred*. Basic Books (2008).

stantiated this in disciplines such as quantum physics, ecology, and systems science. Fritjof Capra, who popularized the surprising parallels between Eastern mysticism and modern physics in his classic 1975 work, *The Tao of Physics*, expresses this so well in his latest book, *The Systems View of Life* (co-authored with Pier Luigi Luisi):

> We have discovered that the material world, ultimately, is a network of inseparable patterns of relationships. We have also discovered that the planet as a whole is a living, self-regulating system. The view of the human body as a machine and of the mind as a separate entity is being replaced by one that sees not only the brain, but also the immune system, the bodily organs, and even each cell as a living, cognitive system. And with the new emphasis on complexity, nonlinearity, and patterns of organization, a new science of qualities is slowly emerging. We call this new science 'the systems view of life' because it involves a new kind of thinking — thinking in terms of relationships, patterns, and context. At the very heart of it, we find a fundamental change of metaphors: from seeing the world as a machine to understanding it as a network.[128]

The change of metaphor from *machine* to *network* is the shift from the *Material* to the *Planetary* consciousness. The machine is the material conception of the world as a collection of independent parts that interact through forces. But the networks of inseparable relationships that comprise Gaia are not material. After all, what is a relationship? It is not an object that you can put on a shelf with a label. Networks of relationships are part of the non-material connective medium that underlies material reality. They are invisible to the five senses and the Material consciousness, but they are visible to the spiritual eyes. Modern science, Eastern mysticism, and indigenous wisdom are coming together as Planetary spirituality.

---

128. *The Systems View of Life*. Cambridge University Press (2014).

The Planetary Human

What do you identify as? Are you male or female? Are you American? Russian? White? Black? Latinx? Are you Christian or Jewish or Islamic? Are you intellectual or emotional? Conservative or liberal? Rich or poor? Ugly or beautiful? A success or a failure? What do all these identity labels mean, and why do we put them on ourselves and others? Is there anything deeper than these categories that we *all* are, some core being-ness we all share?

The answer is, of course, *being human*. We are all having the human experience, with variations on the surface. We all inhabit a body, which itself brings a whole host of issues. We all get hungry and thirsty; we all have strong desires and need connection with other humans. We all need and can give love. We all suffer at times and hopefully experience joy at other times. We are all trying to figure out who we are, and why we get angry and envious. This is the *human condition,* and it exists underneath gender, appearance, nationality, net worth, and anything else we can name. We all have far more in common than we have differences. But the Material consciousness separates and categorizes, creating identities and labels, *us* against *them,* either this or that, *me* versus the world.

Along with the core humanness that we share, we all share this perfect planet. Each and every one of us is part of the interconnected planetary web of life that is Gaia, our living planet. When we begin to recognize, understand, and feel our connectedness with Gaia, we are becoming Planetary humans.

The Planetary consciousness sees the world in connected wholeness, and the Planetary human understands *we* simultaneously with *me,* dissolving the categories and labels of separation. When we have strong identities, such as male, or Buddhist, or white, we are maintaining our separation and closing off our core-beingness, of being *human* first and foremost. Strong identities exclude all those who are *not* in our category, isolating us and limiting our own possibilities.

Mainstream culture today tells me that if I want to be a "real man" I must take charge and have all the answers; I must not cry or be vulnerable or nurturing. But these are culturally conditioned rules that I don't have to accept. In my own experience, when I became a father, it awakened a nurturing and receptive side of myself I never recognized before. It awakened qualities in me that have been defined as "feminine," and as a result I grew into a more complete human being, and not merely a "man." But unfortunately, many men have never found their own feminine qualities; they have suppressed these human qualities because they accept the narrow rules that define what it means to be a man. Women too can discover their "masculine" energy and move beyond the unhealthy rules and limitations of conventional femininity. Once we break away from our conditioned identities and put away the labels, we become capable of the full range of human qualities and of seeing other people as fellow Planetary humans.

We will reclaim more freedom by softening our exclusive identities that keep us boxed in, and awakening to our own core humanness. We can recognize and question the rules of mainstream culture that constrain us, and we can begin to see through the seductive images of pop culture and advertising. Today, while living in "free" democratic societies without authoritarian rule, many people are still held captive by a consumer culture that promotes endless spending and taking on debt to acquire "stuff" in pursuit of happiness.

We idolize and envy pop icons and professional athletes and reality TV stars — all delivered to us through our electronic screens — instead of finding our own freedom and creativity. Our innate talents and creativity are stifled when we try to be like someone else or try to fit the mold of what we are "supposed to be." As we free ourselves from exclusive identities and conditioned roles, we can access our own inner light and bring our unique gifts to the world.

It is the story of separation in the Material consciousness that keeps us confined in identity silos and prevents us from seeing the wholeness of ourselves, and humanity, and life, and the planet. As our consciousness moves from separation to wholeness, we gain more freedom, our creativity is unleashed, and this leads to an increased sense of *agency* — the power to consciously use our creativity to bring intended results. As empowered agents we can take actions that improve the world physically, socially, financially, culturally, spiritually — the possibilities are open.

Humans can choose to use their agency for many purposes — to build a factory that reaps profits but discharges toxic waste into the environment or to create an organization that feeds the hungry. Agency is our birthright and most powerful tool, but it must be used for right purpose. Throughout the last 5,000 years, agency has been limited to those in power, and more recently to the affluent, while everyone else is constrained. But in the future world we want, all of us will awaken into our own agency, and no one will be held down by power structures or cultural conditioning.

The vision and the dream for humans in the Planetary culture of the future is to become free of power structures, social structures, and cultural conditioning, so that each of us can find our *own* calling that contributes to the well-being of other people and the planet. The Planetary human is the empowered agent operating in harmony with the sacred planetary whole.

The Power of Personal Choice

We each carry a personal culture within us that defines who we are — our identities, our values and beliefs, and our worldview. Personal culture governs our actions and our choices. Until recently, personal culture was derived mostly from family and ethnic traditions, but now, for most people, personal culture is determined largely by who we associate with — socially and at work — and what we feed our brains. What most people today are feeding their brains is coming down the big electronic pipeline and straight into the cortex: TV, streaming services, radio, YouTube, Facebook, Twitter, Instagram, and so on.

Let's each of us examine the influences on our own personal culture and ask whether they bring us well-being or harm. Does our media consumption inspire and empower us, or does it make us feel inadequate and envious of the perfect people we see on screen? Do the people we choose to associate with support, encourage, and elevate who we genuinely are, or do they discourage and diminish us? What values are expressed and reinforced within our social environment? For example, if I go out drinking regularly with a group of men who enjoy telling demeaning stories about women, I am supporting a culture of misogyny and assimilating it into my own personal culture. I will inevitably then spread that to others, and perhaps to my children.

Today, culture spreads in many ways. Screen technologies have become very influential, but person-to-person contact is still as important as it was 3 million years ago. We must each recognize our own powerful role in influencing others by our words and actions and consciously choose to be spreaders of positive culture. We are all embedded in culture — we receive it and we transmit it — and we can make conscious personal choices about what we allow into our minds and what we send out to other minds through our actions (including words). When we become more aware of what comes in and what goes out, we can make choices that generate positive culture — within ourselves and out in the world.

Most of us wonder if just one person — little ol' me — can actually make a difference in catalyzing a large-scale shift to positive culture. It's tempting to consider ourselves as just one person out of *eight billion* on the planet, one very tiny cog in the giant machine of civilization. But our individual choices, our actions, our commitments, and our personal culture are what we bring to the world, and if many of us are aligned and work together, it will change the world.

In that spirit I offer the following suggestions for actions and commitments any of us can choose that generate well-being and positive culture. Any of these you already do, or begin doing, are steps in the right direction!

1. **Help stabilize Earth's human population.** This is one of the most important things we need to work on. If you choose to have children, limit that to one or two, and consider adoption as an alternative. Support organizations that educate girls and women and make birth control available. To find out about ongoing efforts to address

population growth, here are two excellent organizations you can learn from and support:

Population Connection: https://www.populationconnection.org/

Population Matters: https://populationmatters.org/

2. **Reduce your footprint on the planet.** To find out more about what contributes to your footprint and to calculate your own footprint go to the *Global Footprint Network*, https://www.footprintnetwork.org/. Here are some specific things you can do to reduce your footprint:

   - **Drive less**. Walk, bike, carpool, ride share, and use public transportation more.

   - **Don't waste energy** in your home or workplace. Improve insulation. Use low consumption, high efficiency appliances and lighting. Dry your laundry outdoors in fresh air whenever you can instead of using a clothes dryer. Don't do silly, wasteful things like running the heat or AC with the windows open.

   - **Use renewable energy** if you are financially able to. Install solar or buy wind energy, or support utilities that are providing energy from renewable sources.

   - **Eat less meat.**

   - **Practice circular economics: Reduce** your consumption of "stuff" you will eventually throw away, and avoid excess packaging. **Reuse** as much as possible (like plastic bags and containers). **Recycle** everything you possibly can.

   - **Buy local** as much as possible. This reduces the carbon footprint from long-distance transportation, and it supports your local economy.

3. **Exercise your buying power** by supporting companies committed to well-being and boycotting companies that are exploitive of labor and nature, or otherwise do harm.

4. **Serve your community** through volunteerism, activism, providing services, helping others, building community networks of support, serving on community boards, tutoring or mentoring younger people, spending time with older people who are alone, or just practicing kindness to everyone.

5. **Support organizations that generate well-being** by starting one, working for one, volunteering, or donating.

6. **Vote and advocate** for candidates and policies that promote justice and well-being for all people.

7. **Create value in the world by doing work that is needed.** One business strategy is to create a product or service and then convince people they need it. Another is to see what is *needed* and then create a product or service that fulfills that need.

8. **Elevate your own consciousness.** The *Seven Markers* in Chapter Six and the one-page summary at the end of that chapter are intended to inspire this work, but once you are on board with the *evolution of consciousness* there are many helpful practices and resources available. Here are just a few suggestions for supporting your own personal awakening and well-being:

   - Develop a daily awareness practice (see *Awakening*, pg 189) that quiets your thinking and stabilizes attention, thus allowing your consciousness to evolve naturally. There are many helpful meditation apps available for mobile phones.

   - Choose to consume inspiring, empowering, positive content that elevates you, whether that be books, magazines, or electronic media.

   - Practice listening. Enjoy face-to-face interaction with people (not just through electronics) and seek quality conversation that balances talking and listening. See also *Community Dialogs* below.

   - Spend time in nature as your living situation allows. Travel softly, listen deeply, and stop often to soak in the restorative energy of Gaia.

- Consciously choose how you use electronic technology, such as computers, smartphones, TV, radio, etc. As with all tools, use them to serve you, for a clear purpose, then *put them away* when they have served that purpose.

- Support your body's optimal functioning by eating healthy and exercising.

- Make time to be alone and quiet, to reflect or contemplate and hear your inner voice.

You probably have more ideas that could be added to this list. Wouldn't it be great if we could share our ideas and our commitments, so we could inspire and support each other? (See *Initiative One* below). Imagine how the world would change if large numbers of people did these things. It starts with you and me.

Large-scale Initiatives for Cultural Transformation

Change can happen through the personal choices we make, but we as a planetary community must also build structures that support and catalyze positive culture on a large scale, across global society. Here I suggest three mechanisms for catalyzing evolutionary change that are possible today and would have a powerful positive influence on global culture. At present, these are ideas in the formative stage and I am working with others to develop them. Maybe you can help.

Initiative One – *Planet:* A new, internet-based platform for generating positive culture.

Connective platforms, also known as social media sites, are proliferating on the internet today, and according to *Search Engine Journal* the seven largest social media sites in 2020 were:

Facebook	2.45 billion users
Instagram	1 billion users

Reddit	430 million
Snapchat	360 million
Twitter	330 million
Pinterest	322 million
LinkedIn	310 million

That is a lot of people engaged in social media — a large fraction of the world's population! We know that the internet and social media platforms can reach billions of people, something never before possible. This is probably the most powerful influence on people and culture today, or at any time in the past. But how wisely are we using these tools? What benefits are they bringing to people? Do they generate well-being in the world?

Certainly, today's social media platforms can do some good. They allow us to communicate and connect with others and share news, information, and ideas. They support e-commerce, which is good for business and can help us acquire things we need. There are probably other positives but, overall, how much well-being do these platforms bring into the world, and how much harm? Social media platforms are also being used to promote hate, violence, white nationalism, and baseless conspiracy theories, and there is mounting evidence that excessive use of electronic media can be detrimental to health and well-being in many ways. As incredible and wonderful as our digital technologies may seem, we are not yet using them very wisely or beneficially. So far, the much-vaunted age of information leaves much to be desired.

What if we could start using these technologies for the right purposes, to promote *well-being* on the planet? What if we had a new social media platform dedicated to benefiting humanity and building positive culture, instead of just generating profits for a few people or manipulating people's beliefs? It doesn't exist yet, but until someone builds it and gives it a better name let's call it *Planet*, a post-Facebook connective platform that supports Planetary culture. Here, like-minded people would be able to access content that is positive and empowering, to find each other and support each other in personal and collective evolution by sharing positive narratives and resources that inspire and uplift.

Imagine *Planet* as a global community of people with the common vision of a world that works for everyone. *Planet* could eventually do what Facebook and other platforms do, but without the harmful and exploitive aspects, such as dis-information and personal data mining. *Planet* could be carefully moderated to allow only content that generates well-being and contributes to the evolution of consciousness — no hate, fear, or xenophobia. *Planet* would broadcast what we are *for* not what we are *against*. It could include forums for civil discourse, and it could be a repository for educational media, videos, music, and other resources that promote personal evolution and elevate the community of users. When *Planet* has billions of users it would become a transformative force for spreading Planetary culture. With the right resources and the right people involved, this could happen very rapidly.

<u>Initiative Two</u> – *The Planetary Culture Collective:* A positive culture factory.

This initiative would create a collaborative media production entity that produces and broadcasts state-of-the-art media of all types that is positive, inspiring, and uplifting, promoting the well-being of people and planet. We'll call this entity the *Planetary Culture Collective* for now. Since there are already companies and organizations doing this,[129] the Planetary Culture Collective will also act as an umbrella that connects and supports aligned efforts, by providing and sharing resources and crafting unified messages. Positive media can be attractive and compelling, as was, for example, James

---

129. A few examples are:

Uplift TV, a free online service out of Australia featuring short films and other content that promotes spiritual growth and well-being. www.uplift.tv.

Positive.News, a print and digital magazine out of Great Britain focusing on what went right. www.positive.news.

Yes! Magazine, a print and digital magazine out of the U.S. and founded by David and Fran Korten, dedicated to social justice, environment, and well-being. www.yesmagazine.org.

Gaia, a member supported media network based in the U.S. with over 8,000 ad-free, streaming titles that support the evolution of consciousness. www.gaia.com.

Cameron's film *Avatar*, which broke multiple box office records and reached hundreds of millions of people throughout the world with positive memes.

Already many young artists (and some old ones) are creating positive, inspiring, and empowering media on YouTube, Vimeo, and other platforms. The Planetary Culture Collective could fund contests that gave lucrative awards to the best positive media from freelance artists, and it could provide channels for global distribution and broadcast. World class film makers, musicians, artists, writers, directors, producers, and other creative minds could be attracted into the effort to produce compelling entertainment and art that spreads the narratives of Planetary culture. The positive media produced through the Collective would give audiences an attractive alternative to violence, xenophobia, misogyny, racism, and other harmful memes that are widespread today.

Initiative Three – *Community Dialog Circles*

Unlike the first two, this initiative is non-technological, local, and low cost; it creates positive culture from the grassroots. Most people today need more real-person, face-to-face interaction and a community to be part of. This initiative is a simple community practice addressing that need, and it also builds local cultures of connection, healing, inclusive decision-making, and well-being.

A Community Dialog Circle starts as an open experiment that can evolve organically and uniquely according to needs and interests of the community it serves. There is no right or wrong way to do this, and there are no strict rules, except to avoid harm and seek well-being. Here are some suggested steps and structures:

1. Choose a time and a place to meet, such as a public meeting room or a home with enough space.

2. Choose a general theme for the dialog, such as racism in the community, or reducing crime, or addressing homelessness, or making the world a better place, or …

3. Invite 5-10 community members to start with. But who to invite? This is part of the experiment.

4. A skilled facilitator is needed. This person manages the group dynamics with awareness and gentle assertiveness. Skillful facilitation is an art and a science that can be learned, and some people have a natural gift for this.

5. Sit in a circle.

6. Make it a goal to hear from everyone, fairly and equally. This means that each person will be mostly *listening*. In a dialog circle all participants are equal, all voices are respected and heard. The facilitator insures fairness and safety.

7. Many processes can be used, but *circling* is one of the most basic and effective. Go around the circle in clockwise fashion and have each person briefly (about 30 seconds to a minute) express whatever it is that's coming up for them at that moment (typically related to the theme). It can be a response to what an earlier person said or something new. After one round, some sort of pattern, or theme, or wisdom may emerge; or additional rounds can be done. You will be surprised by what comes up from this process.

8. A dialog group may meet only once to work on a one-time issue or on a regular basis. The group may want to keep the same participants or invite new ones. Some dialog groups may have action outcomes, where an agreement is reached on implementation of the ideas expressed. These decisions are up to the group, as the experiment unfolds.

Dialog circles are not new, but community interaction like this has been lost to most of us today. Since the 1960s, many experiments have been carried out with group dialog, ranging from "encounter groups" to AA-style meetings. As more people try this in their own communities, we could share best practices, such as norming, circling, heart connection practices, and effective facilitation, as well as new forms and formats of dialog. This knowledge could be shared and transmitted through the *Planet* platform.

It is not so important what one group accomplishes in their dialog circle. What is most important is that local communities everywhere start employing this method for community connection, communication, and decision-making. Local governments can use this to engage citizens and listen to the communities they serve. Most people today need connection with a community and will probably find this practice quite enjoyable — a new option for Saturday night! This can transform the fabric of our culture from the bottom up.

Through the combination of personal commitments to action, community engagement, and large-scale efforts like the initiatives above, we absolutely *can* transform the culture and consciousness of humanity into the Planetary. This is how we can create an ecological civilization that serves all people, unleashes the creativity and agency of everyone, and honors our sacred planet.

Epilogue

In the Introduction, I stated that this book would center around two questions, so let us return to those and consider what we have found. First, *what are humans, and what is our story?* I have made the case that humans are a form of life unlike any other on the planet, and what defines us uniquely is our evolving consciousness. When our ancestors first began making and using stone tools 3 million year ago, it signaled the beginning of a new era in the evolution of life, the era of knowledge and culture and evolving consciousness.

Our story from that point until now can be told in four chapters: the *Mimetic* culture, the *Magical* culture, the *Mythical* culture, and the *Material* culture. These are not only stages in the evolution of culture that are evidenced in the archaeological record, but they are also structures of evolving consciousness that emerged in succession as the *Homo* lineage evolved, each with its own form of meaning and its own way of seeing the world. These four structures are also the broad landscape of our own consciousness — all of these are still present and active within each of us. And finally, these structures also manifest as life stages we move through from birth.

Our own life is a reenactment of the evolutionary journey of humanity and each of us are capable of evolving beyond the Material conception of the world and into the *Planetary* consciousness.

This deep ancestral history is new for most of us, and it helps us see our modern culture and the events of today as being part of a much bigger unfolding. This new story of humanity tells us that, as impressive as today's technology seems, we are still living in a 5,000 year-old civilizational paradigm — the Material culture — characterized by the exploitation and destruction of nature; by power structures that concentrate wealth and limit freedom; by a belief in the primacy of material objects; and by a pervasive self-centeredness that disconnects us. But as our planet's population swells to eight billion, doubling since Neil Armstrong walked on the Moon in 1969, it has become clear that our Material culture and worldview is not sustainable. We are now surpassing the boundaries of the planetary support system and living on the credit card, which brings us to the second question.

What is the future of humanity? Of course we can only speculate, but one possibility is the outright extinction of *Homo sapiens*, the last remaining member of the *Homo* lineage. I consider this unlikely because we are a resourceful and resilient species. More likely, if we don't change our ways, we face a bleak and dystopic future of pollution and toxic waste, of collapsing ecosystems and crumbling civilization, of chaos and war and violence and pandemics. However, the point of this book is to present an alternative scenario, an evolutionary vision in which humanity moves into the next chapter of our story, the *Planetary* culture, in which all of humanity can thrive in harmony with Gaia.

This may sound like a rose-colored fantasy. But evolution is a powerful and active force in the universe, and the evolution of consciousness is the hallmark of humanity. We have been on an ever-accelerating evolutionary path since the birth of the *Homo* lineage. Our brain tripled in size, and our culture progressed from the simplest stone tools to smart phones and the internet. No other living thing has undergone anything like this rapid and dramatic evolutionary change. Humans are on an explosive evolutionary fast-track! It's not that we have become physically superior and can survive better — we mastered survival long ago. Now the game is the evolution of consciousness, and that is why the future can be so very different.

We're not stuck where we have been for 5,000 years, as the exploiters of a material world, striving only to maximize our own personal gains. We are capable of much more. We are destined for something far greater. A new consciousness is emerging that feels and knows the interconnected wholeness and the sacredness of Gaia. From this consciousness a new culture will follow that supports the well-being of all people and empowers each of us to find and share our true gifts. For all of us who have awakened to the vision of the emerging world, we must find each other and, together, help bring it forth.

Appendices

Appendix I – Suggested Readings

Bohm, David. *Wholeness and the Implicate Order.* Routledge (1980, 1990)

Briggs, Roger P. *Journey to Civilization: The Science of How We Got Here.* Collins Foundation Press (2013)

Brownlee, Donald and Ward, Peter D. *Rare Earth: Why Complex Life is Uncommon in the Universe.* Copernicus (2000)

Bruyere, Rosalyn L. *Wheels of Light: Chakras, Auras, and the Healing Energy of the Body.* Simon and Schuster (1994)

Capra, Fritjof and Luisi, Pier Luigi. *The Systems View of Life: A Unifying Vision.* Cambridge University Press (2014)

Chopra, Deepak and Kafatos, Menas. *You Are the Universe: Discovering Your Cosmic Self and Why it Matters.* Harmony Books (2017)

Damasio, Antonio. *Self Comes to Mind: Constructing the Conscious Brain.* Vintage Books (2010)

Dawkins, Richard. *The Greatest Show on Earth: The Evidence for Evolution.* Free Press (2009)

Donald, Merlin. *Origins of the Modern Mind: Three Stages in the Evolution of Culture and Cognition.* Harvard University Press (1991)

Donald, Merlin. *A Mind So Rare: The Evolution of Human Consciousness.* W.W. Norton (2001)

Eisenstein, Charles. *The More Beautiful World Our Hearts Know is Possible.* North Atlantic Books (2013)

Feuerstein, Georg. *Structures of Consciousness: The Genius of Jean Gebser.* Integral Publishing (1987)

Feuerstein, Georg. *Jean Gebser: What Color is Your Consciousness?* Robert Briggs Associates (1989)

Garfield, Jay L. *Engaging Buddhism.* Oxford University Press (2015)

Gebser, Jean. *The Ever-Present Origin.* Ohio University Press (1985)

Hubbard, Barbara Marx. *Conscious Evolution: Awakening the Power of Our Social Potential.* New World Library (1998)

Johnson, Jeremy. *Seeing Through the World: Jean Gebser and the Integral Consciousness.* Revelore Press (2019)

Kauffman, Stuart. *Reinventing the Sacred: A New View of Science, Reason, and Religion.* Basic Books (2010)

Klein, Richard, with Blake Edgar. *The Dawn of Human Culture.* John Wiley and Sons (2002)

Korten, David. *The Great Turning: From Empire to Earth Community.* Barrett-Koehler (2006)

Korten, David. *Agenda for a New Economy: From Phantom Wealth to Real Wealth.* Barrett-Koehler (2009)

Korten, David. *When Corporations Rule the World.* Barrett-Koehler (2015)

Korten, David. *Change the Story, Change the Future.* Barrett-Koehler (2015)

Laszlo, Ervin. *Science and the Akashic Field: An Integral Theory of Everything.* Inner Traditions (2004)

Laszlo, Ervin. *The Akashic Experience: Science and the Cosmic Memory Field.* Inner Traditions (2000)

Lovins, Hunter, with Stewart Wallis, Anders Wijkman, and John Fullerton. *A Finer Future.* New Society (2018)

McIntosh, Steve. *Integral Consciousness and the Future of Evolution.* Paragon House (2007)

McIntosh, Steve. *Evolution's Purpose.* SelectBooks (2012)

Norris, Zach. *We Keep Us Safe: Building Secure, Just, an Inclusive Communities.* Beacon Press (2020)

Raworth, Kate. *Doughnut Economics: Seven Ways to Think Like a 21st Century Economist.* Chelsea Green (2017)

Richerson, Peter J. and Boyd, Robert. *Not By Genes Alone: How Culture Transformed Human Evolution.* University of Chicago Press (2005)

Rifkin, Jeremy. *The Empathic Civilization: The Race to Global Consciousness in a World in Crisis.* Tarcher/Penguin (2009)

Rockström, Johan, and Klum, Mattias. *Big World, Small Planet.* Max Strom Publishing (2015)

Schleicher, Joan (editor). *The Theory of Process 1 and 2.* Robert Briggs Associates (1991)

Shiva, Vandana. *Soil Not Oil: Environmental Justice in an Age of Climate Crisis.* North Atlantic Books (2008)

Stapp, Henry P. *Mindful Universe: Quantum Mechanics and the Participating Observer.* Springer (2007)

Thompson, Evan. *Waking Dreaming, Being: Self and Consciousness in Neuroscience, Meditation and Philosophy.* Columbia University Press (2015)

Wilber, Ken. *The Marriage of Sense and Soul.* Broadway Books (1998)

Young, Arthur M. *The Reflexive Universe: Evolution of Consciousness.* Anodos Publications (1999). First Edition 1976

Young, Arthur M. *The Geometry of Meaning.* Anodos Publications (2011). First Edition 1976

Yunkaporta, Tyson. *Sand Talk: How Indigenous Thinking Can Save the World.* HarperOne (2020)

Appendix II – Key Descriptors Summarizing Gebser's Structures and Stages

Archaic	Magical	Mythical	Mental	Integral
Zero-dimensional	One-dimensional	Two-dimensional	Three-dimensional	Four-dimensional
Timeless	Daily time	Calendric time	Linear time	Time-freedom
Deep Sleep-like	Sleep-like	Dream-like	Wakeful	Diaphanous (transparent)
Instinct	Emotion	Imagination	Cognition	Verition (knowing truth)
Oneness	Idols	Deities	God	Divinity
Origin	Magic	Destiny	Causality	Acausal wholeness
Survival	Ritual	Mythology	Philosophy	Eteology (being-in-truth)
No Self	Clan Self	Tribal Self	Egoic Self	Transparent Self
-	Pre-rational	Irrational	Rational	Arational
Unitary	Non-causal	Pre-causal	Causal	Acausal
Non-duality	Non-perspectival	Pre-perspectival	Perspectival	Aperspectival

Appendix III – Diving Deeper into Arthur Young: The Torus and Seven-ness

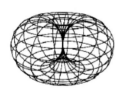

Courtesy: Institute for the Study of Consciousness

The V-shaped arc of process is a two dimensional depiction of evolution, but in three dimensions this evolutionary movement can be illustrated by a donut shape that mathematicians call a *torus* (left). This shape has many unique properties; one is that the center and the surface are completely connected. For a sphere or any other closed surface, the center and the surface are completely separate. The torus can be seen in nature in the shape of magnetic fields and in the circulation vortices of fluids, such as in tornados and swirling water.

Young concluded that "the universe and the creatures which inhabit it are toroidal." What does this mean? When he was beginning his work on the theory of process, Young questioned some aspects of Einstein's general theory of relativity that reimagined gravity as curved space. This curvature could be either spherical, where parallel lines converge, or saddle-like, where parallel lines diverge (general relativity shot down the Euclidean space of basic geometry where parallel lines never meet). The great physicist and mathematician Arthur Eddington was also working on models of the universe with Einstein and they proposed a model with saddle-like curvature. They called this the *hypersphere*, and it had a volume given by $2\pi^2 R^3$. Young realized that $2\pi^2 R^3$ was also the volume of a torus with an infinitely small hole, something that Einstein, Eddington, and others apparently did not recognize. Because the torus and the hypersphere have the same volume formula, Young concluded that the universe must be toroidal. In 1974 John Archibald Wheeler, one of the great cosmologists of the last century, reached the same conclusion.

The torus also contributed to Young's understanding of *seven* as the number of stages in evolutionary processes. First, he noticed seven-ness in many of the world's mythologies and wisdom traditions. In the Judeo-Christian version, God made the world in seven days, and this same idea is found in Zoroastrian and Japanese creation mythologies. Seven-ness is also prominent in Greek mythology (such as the myth of Cronus) and is central in ancient Hindu traditions, such as the seven chakras of the body. The recurrent theme of seven-ness throughout ancient traditions prompted Young to explore this further through mathematics.

Young gives a mathematical "proof" of seven-ness in Appendix II of *The Reflexive Universe*, but you'll need to be a bona fide mathematician to fully appreciate the arguments (I'm not). However, another answer comes from the mathematical field of topology and the famous "colored map problem." The problem can be stated this way: What is the *least* number of colors you would need to make a map of many regions, with no regions of the same color touching each other? We could ask this question for a map of the 48 states in the continental U.S. As simple as it sounds this problem took more than 100 years to solve, and it was finally proven in 1976 that the answer is *four* colors. This is now called the *Four-color theorem*, and it applies to flat maps and also to spheres and any other closed surfaces — except the torus. If we ask the colored map question for a torus instead of a plane or sphere, the answer is *seven*.

Courtesy: David Sibbet

Young saw that seven colored regions can be laid out on the surface of a torus, one after another, along a spiral that begins at the center and moves out the top in a left-handed turn until reaching the farthest out point (at the far right edge of the torus shown). The spiral continues around the bottom of the torus to enter the bottom and return to its starting point at the center.

In Young's vision, the seven stages of process occur on the surface of a torus, one after another, following the spiral and spreading out to cover the entire surface. Stage 4 of process, with the Turn, is the point furthest out on the spiral where it begins to fall back inward.

Appendix IV – Summary of the Gebser-Donald-Young Macro-stages

~ Date of Emergence	Macro-stage	Key Happenings	Anthropology
3 million years ago	**Mimetic** Gesture, skill, embodied enactment, mime	**FIRST STONE TOOLS** Beginning of semantic culture and trans-generational knowledge	Australopithecus Africanus, Homo habilis, Homo rudolfensis, or ?
2 million years ago	Nuclear Families		Homo ergaster (Africa), Homo erectus (Asia)
1 million years ago	**Magical** Nature spirits, ceremony, shamanism, separation from nature	**ELABORATE RITUAL** Large game hunting, control of fire, porto-language, simple shelters	Homo heidelbergensis (Africa and, later, Europe) Homo neanderthalensis, (Eurasia)
300,000 years ago 70,000 years ago	Clans	Earliest known Homo sapiens Near-extinction of humans	Homo sapiens idaltu (archaic humans in Africa)
60,000 years ago	**Mythic** Oral-mythical culture, dream-like reality, tribal self Tribes	**THE GREAT LEAP** Tool explosion, mastery of spoken language, final diaspora out of Africa	Homo sapiens sapiens (behaviorally modern humans) Cro-Magnon culture, Cave Paintings
5,000 years ago	**Material** World as objects-in-space, separation, causality	**CIVILIZATION** Written language, top-down power structures, city states and empires	The built world, monumental architecture, hierarchical societies
2,500 years ago	Mental Theoretic Consciousness Egoic self	The Axial Age	Classical Greece, the "modern mind"
400 years ago	Mental Rational Consciousness Nations	Scientific revolution	Classical physics, industrial revolution
60 years ago		The Counterculture	Post-modernism, the Beatles
Near Future	**Planetary** Planetary spirituality, Integral consciousness	**PLANETARY CULTURE** Wellbeing economics and ecological civilization	Homo sapiens planetus? (Planetary humans)

Appendix V – The Earth Charter

ORIGIN OF THE EARTH CHARTER

The Earth Charter was created by the independent Earth Charter Commission, which was convened as a follow-up to the 1992 Earth Summit in order to produce a global consensus statement of values and principles for a sustainable future. The document was developed over nearly a decade, through an extensive process of international consultation, to which over five thousand people contributed. The Charter has been formally endorsed by thousands of organizations, including UNESCO and the IUCN (World Conservation Union). For more information, please visit www.EarthCharter.org.

PREAMBLE

We stand at a critical moment in Earth's history, a time when humanity must choose its future. As the world becomes increasingly interdependent and fragile, the future at once holds great peril and great promise. To move forward we must recognize that in the midst of a magnificent diversity of cultures and life forms, we are one human family and one Earth community with a common destiny. We must join together to bring forth a sustainable global society founded on respect for nature, universal human rights, economic justice, and a culture of peace. Towards this end, it is imperative that we, the peoples of Earth, declare our responsibility to one another, to the greater community of life, and to future generations.

Earth, Our Home

Humanity is part of a vast evolving universe. Earth, our home, is alive with a unique community of life. The forces of nature make existence a demanding and uncertain adventure, but Earth has provided the conditions essential to life's evolution. The resilience of the community of life and the well-being of humanity depend upon preserving a healthy biosphere with all its ecological systems, a rich variety of plants and animals, fertile soils, pure waters, and clean air. The global environment with its finite resources is a common concern of all peoples. The protection of Earth's vitality, diversity, and beauty is a sacred trust.

The Global Situation

The dominant patterns of production and consumption are causing environmental devastation, the depletion of resources, and a massive extinction of species. Communities are being undermined. The benefits of development are not shared equitably, and the gap between rich and poor is widening. Injustice, poverty, ignorance, and violent conflict are widespread and are the cause of great suffering. An unprecedented rise in human population has overburdened ecological and social systems. The foundations of global security are threatened. These trends are perilous — but not inevitable.

The Challenges Ahead

The choice is ours: form a global partnership to care for Earth and one another or risk the destruction of ourselves and the diversity of life. Fundamental changes are needed in our values, institutions, and ways of living. We must realize that when basic needs have been met, human development is primarily about being more, not having more. We have the knowledge and technology to provide for all and to reduce our impacts on the environment. The emergence of a global civil society is creating new opportunities to build a democratic and humane world. Our environmental, economic, political, social, and spiritual challenges are interconnected, and together we can forge inclusive solutions.

Universal Responsibility

To realize these aspirations, we must decide to live with a sense of universal responsibility, identifying ourselves with the whole Earth community, as well as our local communities. We are at once citizens of different nations and of one world in which the local and global are linked. Everyone shares responsibility for the present and future well-being of the human family and the larger living world. The spirit of human solidarity and kinship with all life is strengthened when we live with reverence for the mystery of being, gratitude for the gift of life, and humility regarding the human place in nature.

We urgently need a shared vision of basic values to provide an ethical foundation for the emerging world community. Therefore, together in hope we affirm the following interdependent principles for a sustainable way of life as a common standard by which the conduct of all individuals, organizations, businesses, governments, and transnational institutions is to be guided and assessed.

PRINCIPLES

PILLAR I: RESPECT AND CARE FOR THE COMMUNITY OF LIFE

1. Respect Earth and life in all its diversity.

  a. Recognize that all beings are interdependent and every form of life has value regardless of its worth to human beings.

  b. Affirm faith in the inherent dignity of all human beings and in the intellectual, artistic, ethical, and spiritual potential of humanity.

2. Care for the community of life with understanding, compassion, and love.

  a. Accept that with the right to own, manage, and use natural resources comes the duty to prevent environmental harm and to protect the rights of people.

  b. Affirm that with increased freedom, knowledge, and power comes increased responsibility to promote the common good.

3. Build democratic societies that are just, participatory, sustainable, and peaceful.

  a. Ensure that communities at all levels guarantee human rights and fundamental freedoms and provide everyone an opportunity to realize his or her full potential.

  b. Promote social and economic justice, enabling all to achieve a secure and meaningful livelihood that is ecologically responsible.

4. Secure Earth's bounty and beauty for present and future generations.

  a. Recognize that the freedom of action of each generation is qualified by the needs of future generations.

b. Transmit to future generations values, traditions, and institutions that support the long-term flourishing of Earth's human and ecological communities. In order to fulfill these four broad commitments, it is necessary to:

(continue to Pillar II)

PILLAR II: ECOLOGICAL INTEGRITY

5. Protect and restore the integrity of Earth's ecological systems, with special concern for biological diversity and the natural processes that sustain life.

    a. Adopt at all levels sustainable development plans and regulations that make environmental conservation and rehabilitation integral to all development initiatives.

    b. Establish and safeguard viable nature and biosphere reserves, including wild lands and marine areas, to protect Earth's life support systems, maintain biodiversity, and preserve our natural heritage.

    c. Promote the recovery of endangered species and ecosystems.

    d. Control and eradicate non-native or genetically modified organisms harmful to native species and the environment, and prevent introduction of such harmful organisms.

    e. Manage the use of renewable resources, such as water, soil, forest products, and marine life, in ways that do not exceed rates of regeneration and that protect the health of ecosystems.

    f. Manage the extraction and use of non-renewable resources, such as minerals and fossil fuels, in ways that minimize depletion and cause no serious environmental damage.

6. Prevent harm as the best method of environmental protection and, when knowledge is limited, apply a precautionary approach.

    a. Take action to avoid the possibility of serious or irreversible environmental harm even when scientific knowledge is incomplete or inconclusive.

b. Place the burden of proof on those who argue that a proposed activity will not cause significant harm and make the responsible parties liable for environmental harm.

c. Ensure that decision making addresses the cumulative, long-term, indirect, long distance, and global consequences of human activities.

d. Prevent pollution of any part of the environment and allow no build-up of radioactive, toxic, or other hazardous substances.

e. Avoid military activities damaging to the environment.

7. **Adopt patterns of production, consumption, and reproduction that safeguard Earth's regenerative capacities, human rights, and community well-being.**

    a. Reduce, reuse, and recycle the materials used in production and consumption systems, and ensure that residual waste can be assimilated by ecological systems.

    b. Act with restraint and efficiency when using energy, and rely increasingly on renewable energy sources, such as solar and wind.

    c. Promote the development, adoption, and equitable transfer of environmentally sound technologies.

    d. Internalize the full environmental and social costs of goods and services in the selling price, and enable consumers to identify products that meet the highest social and environmental standards.

    e. Ensure universal access to health care that fosters reproductive health and responsible reproduction.

    f. Adopt lifestyles that emphasize the quality of life and material sufficiency in a finite world.

8. **Advance the study of ecological sustainability and promote the open exchange and wide application of the knowledge acquired.**

    a. Support international scientific and technical cooperation on sustainability, with special attention to the needs of developing nations.

b. Recognize and preserve the traditional knowledge and spiritual wisdom in all cultures that contribute to environmental protection and human well-being.

c. Ensure that information of vital importance to human health and environmental protection, including genetic information, remains available in the public domain.

PILLAR III: SOCIAL AND ECONOMIC JUSTICE

9. Eradicate poverty as an ethical, social, and environmental imperative.

  a. Guarantee the right to potable water, clean air, food security, uncontaminated soil, shelter, and safe sanitation, allocating the national and international resources required.

  b. Empower every human being with the education and resources to secure a sustainable livelihood, and provide social security and safety nets for those who are unable to support themselves.

  c. Recognize the ignored, protect the vulnerable, serve those who suffer, and enable them to develop their capacities and to pursue their aspirations.

10. Ensure that economic activities and institutions at all levels promote human development in an equitable and sustainable manner.

  a. Promote the equitable distribution of wealth within nations and among nations.

  b. Enhance the intellectual, financial, technical, and social resources of developing nations, and relieve them of onerous international debt.

  c. Ensure that all trade supports sustainable resource use, environmental protection, and progressive labor standards.

  d. Require multinational corporations and international financial organizations to act transparently in the public good and hold them accountable for the consequences of their activities.

11. **Affirm gender equality and equity as prerequisites to sustainable development and ensure universal access to education, health care, and economic opportunity.**

    a. Secure the human rights of women and girls and end all violence against them.

    b. Promote the active participation of women in all aspects of economic, political, civil, social, and cultural life as full and equal partners, decision makers, leaders, and beneficiaries.

    c. Strengthen families and ensure the safety and loving nurture of all family members.

12. **Uphold the right of all, without discrimination, to a natural and social environment supportive of human dignity, bodily health, and spiritual well-being, with special attention to the rights of indigenous peoples and minorities.**

    a. Eliminate discrimination in all its forms, such as that based on race, color, sex, sexual orientation, religion, language, and national, ethnic, or social origin.

    b. Affirm the right of indigenous peoples to their spirituality, knowledge, lands and resources and to their related practice of sustainable livelihoods.

    c. Honor and support the young people of our communities, enabling them to fulfill their essential role in creating sustainable societies.

    d. Protect and restore outstanding places of cultural and spiritual significance.

PILLAR IV: DEMOCRACY, NONVIOLENCE, AND PEACE

13. **Strengthen democratic institutions at all levels, and provide transparency and accountability in governance, inclusive participation in decision making, and access to justice.**

    a. Uphold the right of everyone to receive clear and timely information on environmental matters and all development plans and activities which are likely to affect them or in which they have an interest.

   b. Support local, regional, and global civil society, and promote the meaningful participation of all interested individuals and organizations in decision making.

   c. Protect the rights to freedom of opinion, expression, peaceful assembly, association, and dissent.

   d. Institute effective and efficient access to administrative and independent judicial procedures, including remedies and redress for environmental harm and the threat of such harm.

   e. Eliminate corruption in all public and private institutions.

   f. Strengthen local communities, enabling them to care for their environments, and assign environmental responsibilities to the levels of government where they can be carried out most effectively.

14. **Integrate into formal education and life-long learning the knowledge, values, and skills needed for a sustainable way of life.**

   a. Provide all, especially children and youth, with educational opportunities that empower them to contribute actively to sustainable development.

   b. Promote the contribution of the arts and humanities, as well as the sciences, in sustainability education.

   c. Enhance the role of the mass media in raising awareness of ecological and social challenges.

   d. Recognize the importance of moral and spiritual education for sustainable living.

15. **Treat all living beings with respect and consideration.**

   a. Prevent cruelty to animals kept in human societies and protect them from suffering.

b. Protect wild animals from methods of hunting, trapping, and fishing that cause extreme, prolonged, or avoidable suffering.

   c. Avoid or eliminate to the full extent possible the taking or destruction of non-targeted species.

16. Promote a culture of tolerance, nonviolence, and peace.

   a. Encourage and support mutual understanding, solidarity, and cooperation among all peoples and within and among nations.

   b. Implement comprehensive strategies to prevent violent conflict and use collaborative problem solving to manage and resolve environmental conflicts and other disputes.

   c. Demilitarize national security systems to the level of a non-provocative defense posture, and convert military resources to peaceful purposes, including ecological restoration.

   d. Eliminate nuclear, biological, and toxic weapons and other weapons of mass destruction.

   e. Ensure that the use of orbital and outer space supports environmental protection and peace.

   f. Recognize that peace is the wholeness created by right relationships with oneself, other persons, other cultures, other life, Earth, and the larger whole of which all are a part.

THE WAY FORWARD

As never before in history, common destiny beckons us to seek a new beginning. Such renewal is the promise of these Earth Charter principles. To fulfill this promise, we must commit ourselves to adopt and promote the values and objectives of the Charter.

This requires a change of mind and heart. It requires a new sense of global interdependence and universal responsibility. We must imaginatively develop and apply the vision of a sustainable way of life locally, nationally, regionally, and globally. Our cultural diversity is a precious heritage, and different cultures will find their own distinctive ways to realize the vision. We must deepen and expand the global dialogue that generated the Earth Charter, for we have much to learn from the ongoing collaborative search for truth and wisdom.

Life often involves tensions between important values. This can mean difficult choices. However, we must find ways to harmonize diversity with unity, the exercise of freedom with the common good, short-term objectives with long-term goals. Every individual, family, organization, and community has a vital role to play. The arts, sciences, religions, educational institutions, media, businesses, nongovernmental organizations, and governments are all called to offer creative leadership. The partnership of government, civil society, and business is essential for effective governance.

In order to build a sustainable global community, the nations of the world must renew their commitment to the United Nations, fulfill their obligations under existing international agreements, and support the implementation of Earth Charter principles with an international legally binding instrument on environment and development.

Let ours be a time remembered for the awakening of a new reverence for life, the firm resolve to achieve sustainability, the quickening of the struggle for justice and peace, and the joyful celebration of life.

Index

7th Kingdom 140, 141, 144, 148

A Mind So Rare 60, 117, 123, 126, 277
Acheulean tools 37, 38
Age of the universe 21, 23
Agency 18, 53, 68, 134, 140, 145, 265, 274
AI 57, 66, 73, 74
Akashic Field 198, 199, 220, 278
Alpert, Richard 60
Ambrose, Stanley 41
Anatomically modern humans 9, 39
Arc of process 133-135, 137, 139, 141, 143, 147, 148, 154, 280
Artema, Jelle 73
Artificial Intelligence 57, 66, 108
Aura 81-83
Aurobindo, Sri 160, 187
Australopithecus 32, 34, 35, 100
    Afarensis 35
    Africanus 35, 100
Autobiographical Memory systems 119, 124
Axial Age 93, 96, 107, 120, 123, 125, 150, 164-166

Bardo 80, 81
Beck, Don 92, 97, 129
Behaviorally modern humans 9, 39, 42
Bell model 47 helicopter 132
Bessel, Wilhelm 24
Big Bang cosmology 20, 22
Binding 79, 292
Biofield 84
Black dwarf 25
Black hole 26
Bohm, David 174, 193, 211
Bohr, Niels 69, 174
Boyd, Robert 31
Brown, Scott 194
Bruyere, Rosalyn 82
Buddha, the 76, 93, 153, 191, 259

Cambrian Explosion 8, 9
Capra, Fritjof 18, 262
Cartesian Dualism 55
Çatalhüyük 46, 162
Causality 104, 108, 109, 112, 113, 134, 138, 145, 146, 150, 175-177, 180, 185, 195, 196, 248, 280
Cerebral cortex 63, 64, 86
Chakra 81-84
Chalmers, David 57, 66
Chauvet Cave 42
Circular economics 237, 267
Classical physics 17, 68, 69, 72, 104, 138, 167, 171, 172, 174, 176-178, 180, 181

Club of Rome vi, 228, 230, 238, 241, 242, 246
Co-evolution 122, 160
Collins, Jim 209, 298
Community Dialog Circles 272
Complex eye 14-16
Conscious Capacity 60, 85, 89-91, 105, 107, 189
Consciousness i, iv, 3, 18, 42, 49-51, 53-58, 60, 62-66, 68-79, 81-87, 89-112, 114, 115, 117, 122-132, 144, 149-155, 160-162, 164, 165, 167, 168, 170-173, 176-183, 185-193, 195, 197, 198, 200, 201, 203, 205-207, 209-216, 220-222, 224, 225, 227, 228, 246, 247, 249, 258-264, 268, 271, 274-280, 298, CH. 2
  Access (a-) 57
  Aperspectival Consciousness 111-112, 200-201, 203
  Archaic Consciousness 97, 99, 104
  defintions 54
  Episodic consciousness 123
  in Buddhism iv, 75
  in Neuroscience iv, 60, 62, 279
  in Philosophy iv, 54
  in Physics iv, 68
  in Psychology iv, 58
  Integral Consciousness 94, 97, 111, 114, 129, 152-153, 160, 198, 200, 212, 278, 279
  Magical Consciousness 96, 100-101, 102, 104, 105, Ch. 3-4
  markers Ch. 6
  Material Consciousness 150-152, 161, 162, 165, 170, 176-180, 197, 225, 246, 258, 259, 261-264, Ch. 5
  Mental Consciousness 96, 106-112, 114, 150-151, 165, 168, 171-173, 176, 177, 191, 201, 203, 216, 220, 221, 260
  Mental rational Consciousness 165-166
  Mimetic consciousness 121-130
  Mythic Consciousness 124-126
  Mythical Consciousness 96, 98, 102-107, 110, 112, 128, 150, 249
  outside the brain iv, 81
  Phenomenal Consciousness 57
  Planetary Consciousness 3, 153, 160, 180, 182, 187, 188, 200, 211, 213-215, 225, 228, 246, 247, 262, 263, 275, Ch. 5-7
  Rational Consciousness 108, 111, 165, 167
  structures 84, 107, 116, Ch. 3
  Theoretic Consciousness 109, 125-126, 165, 176
Contingent evolution 13, 14
Convergent evolution 13, 14
Copenhagen Interpretation 71
Cortes, Hernan 110
Cortex 14, 59, 63-65, 67, 68, 85, 86, 265
Cosmic evolution 19, 20, 22, 23, 28, 29, 223
Cosmological constant 195
Counterculture 2, 60, 151, 165, 166, 179
Creative Emergence iv, 17, 18
Cubist period 203
Culture iv, vi, 3, 18, 31-34, 37, 42, 45, 48-51, 61, 90, 91, 93, 95, 97, 98, 100, 103, 105-107, 117, 118, 120-130, 132, 149-151, 153-155, 159-162, 164, 165, 167, 168, 170, 171, 173, 178-182, 187, 193, 204-206, 210, 212, 222-225, 228, 242, 245, 246, 248-251, 253, 254, 256,

258, 260, 261, 264-266, 269-272, 274-279, 283, 291
Animal culture 31, 33
Cro-Magnon culture 105
definition of 31
Episodic culture 33, 120-122, Ch. 4
Material culture 150, 164, 165, 167, 170, 171, 179, 181, 206, 250, 256, 258, 260, 274, 275, Ch. 4-5
Mimetic culture 32, 33, 37, 120-124, 274, Ch. 4
Mythic culture 45, 102, 125, Ch. 3-4
Planetary culture 3, 153, 155, 160-161, 225, 248-250, 256, 258, 265, 270-272, 275, Ch. 6-7

Dalai Lama, the 81
Damasio, Antonio 54, 62-68, 86, 206
Dark energy 23, 194, 195
Dark matter 23, 194
Darwin, Charles 10
Darwinian evolution 11, 12, 17
Dennett, Daniel 56
Descartes, Rene 55
Descent with modification 11
Determinism 16, 133, 134, 138, 139, 145, 146, 248
Diamond, Jared 42
Diaphaneity 113, 212
Diaphanous 113, 114, 280
Dixon-Decleve, Sandrine 242
Doctrine of Discovery 251, 252
Dominion 140, 144
Donald, Merlin v, 3, 33, 48, 60, 109, 115, 117, 118, 122, 123, 125, 126, 129, 130, 149, 152, 160, 297, Ch. 4

Doughnut Economics vi, 230, 231, 237, 247, 279
Dualism 55, 56, 110

Earth Charter vi, 228, 244-246, 283, 291, 292, App. V
Ecological Civilization vi, 227, 228, 237-239, 241, 243, 244, 246, 248, 249, 274
Ego transparency 202, 207
Einstein, Albert 69, 131, 172, 259
Eisler, Riane 161
Emerging New Civilization Initiative 242
Emotional intelligence 216
Enlightenment 106, 107, 151, 153, 191
Entanglement 74, 175, 180, 195-197
Entrainment 217
Epigenetics 16
Epiphenomenalism 62
Episodic Memory systems 119, 124
Eridu 46
Eteology 113, 280
Evolution i, v, 3, 7-20, 22-24, 28-35, 49-51, 56, 60, 64, 75, 86, 87, 89-92, 94-99, 106, 115, 117, 121-123, 125-127, 129-141, 144, 145, 147-154, 159-162, 166, 168, 169, 178, 183, 185-187, 189, 192, 204-206, 210, 222-226, 228, 247, 268, 271, 274-280, 283
Fractal v, 141, 166
of Consciousness i, ii, 3, 51, 86, 87, 91, 92, 95, 97, 99, 106, 115, 122, 123, 126, 160, 178, 186, 187, 210, 224, 225, 247, 268, 271, 275, 276, 279, Ch. 3-7
of Cultures 31
of Life iv, 7-9, 12, 13, 16, 18, 30, 32, 49, 89, 92, 132, 134, 274
of the Self 125, 204, 205, 210

of the Universe iv, 19, 24, 135, 137, 139, 144, 150, 159, 183, 222
Supra-49, 131, 132, 159, 222
Evolutionary Worldview vi, 188, 222-225
Executive Suite 60
Exoplanets 29, 30
Explicate order 176, 180, 193, 194, 196, 221, 261

Feuerstein, Georg 99, 167
First cause 136
fMRI 63
Freud, Sigmund 58, 92, 204

Gage, Phineas 58, 59
Gaia 187, 249, 261-263, 268, 271, 275, 276
Galaxies 19-23, 27
Gebser, Jean iv, 3, 54, 92, 99, 117, 126, 129, 131, 149, 152, 160, 178, 187, 205, 278, Ch. 3
General Theory of Relativity 173
Glial cells 64
Global Footprint Network 214, 267
Göbekli Tepe 45, 161, 162
Goleman, Daniel 216
Gould, Stephen 14
Great transformations iv, 47-49

Hameroff, Stuart 72, 73
Hand axe 42
Hanh, Thich Nhat 193
Hard Problem 58, 66
Heart Coherence 218
Heart Opening vi, 188, 215
Heart Rate Variability 216
HeartMath Institute 85, 216-220, 298
Heisenberg, Werner 69, 70, 93, 174
Hominin 32-34, 37, 48

Homo iv, 3, 8, 9, 31-35, 37-39, 41-43, 45, 47, 48, 64, 91, 95, 100, 102, 105, 106, 117, 120, 121, 123, 128, 140, 144, 148, 149, 153, 155, 159, 210, 224, 274, 275
Erectus 32, 34, 37, 38, 41, 95, 123
Ergaster 38
Genus 33-35
Habilis 34, 37, 38, 95, 100, 123
Heidelbergensis 38, 102, 128
Neanderthalensis 39, 91
Sapiens 8, 9, 31, 32, 34, 39, 41, 42, 91, 102, 105, 106, 128, 275
Sapiens idaltu 39
Sapiens sapiens 42
Homunculus 67
HRV 216, 217
Hubble, Edwin 20
Hubble's constant 21
Hunt, Valerie 82

Idealism 56, 69
Imperial civilization 161
Implicate order 174, 176, 177, 180, 193, 195, 211, 212, 221, 277
Integrated information theory 57
Interbeing 193, 194
Intuition 85, 86, 180-182, 215, 220-222

Jericho 46, 162
Johnson, Jeremy 100, 110, 114
Jung, Carl 92, 93

Kant, Immanuel 56
Kanyini 193
Kanzi 118
Kauffman, Stuart 18, 261
Kingdoms v, 134, 135, 139-141, 144, 148, 183
Klein, Richard 42

Koch, Christof 57
Korten, David 161, 228, 238, 241, 242, 246, 298

Laszlo, Ervin 198, 199
Late Stone Age 42
Leary, Timothy 60
Lenski, Richard 16
Level 5 leader 209, 210
Long-term Evolutionary Experiment 15
Lucid dreaming 78
Lycan, William 58

Macro-stages iii, v, vi, 117, 128-129, 130, 148, 149, 151-155, 160, 166, 209, 224, 246, 282
　Gebser-Donald 128-129
　Gebser-Donald-Young 148-155
　summary 152-155
Manifest Destiny 253
Maslow, Abraham 92, 98, 190, 207, 259
Mass extinction 257
Materialist view 136
McCraty, Rollin 85
Mechanical universe 17, 108, 167, 171, 176, 177
Megalithic structures 45, 161-163, 199
Memory systems 119-121
　Procedural Memory systems 119
Meta-perspective v, 188, 200-203, 213
Microtubules 73-75
Microwave background radiation 22
Mimesis 37, 48, 50, 118, 121-123, 127
Mindfulness 192, 259
Mitchell, Edgar 201
Monad 134, 146
Monism 56
Monkey mind 191, 213
Mont Pelerin Society 229

Morse Simon Conway 14
Mount Toba 41
Mousterian tools 40
Mutations 11, 12, 18, 136

Natural Selection 11, 12, 18, 105
Neanderthals and Humans 39, 40, 44, 102, 106
Neurons 64-66, 73-75, 85, 197
Neutron star 26
Newton, Isaac 17, 68
Newtonian physics 108
Non-locality 175, 180, 195-198, 220, 221
Norris, Zach 255
Nuclear fusion 24, 25

Objective reduction (OR) 74
Observer effect 71
Of the self 56, 67, 79, 80, 112, 125, 204, 205, 210
Olduwan tools 36
Orchestrated OR 75
Origin of elements 26
Origins of the Modern Mind v, 117, 123, 277

Pair creation 137, 140
Pan-psychism 57
Pauli, Wolfgang 69, 174
Penrose, Roger 72, 73
Phenomenology 95
Photoelectric Effect 172, 173
Physicalism 56
Picasso, Pablo 93, 203
Planck, Max 69, 172
Planet 1-3, 7, 9, 18, 19, 23, 25, 27, 30, 43, 57, 90, 103, 105, 115, 124, 140, 144, 152, 160, 167, 170, 179, 182, 184, 186, 200, 210, 211, 213, 214, 223-225,

227-231, 236, 237, 239, 240, 244, 246, 248, 249, 256-258, 260-267, 269-271, 273-275, 279
Planetary v, vi, 2, 3, 25, 26, 29, 129, 145, 153, 155, 160, 161, 166, 170, 180-182, 187, 188, 200, 210, 211, 213-215, 225-228, 231-235, 237, 242, 243, 246-250, 256, 258, 260-265, 269-272, 274, 275
   Boundaries 227, 231, 233-235
   Consciousness 3, 153, 160, 180, 182, 187, 188, 200, 211, 213-215, 225, 228, 246, 247, 262, 263, 275, Ch. 5-7
   Culture 3, 153, 155, 160-161, 225, 248-250, 256, 258, 265, 270-272, 275, Ch. 6-7
   Human vi, 153, 260, 263-265
   Spirituality 258-263
Planetary Emergency Initiative 242, 243
Planetary systems 26, 29, 231
Planets 19, 20, 23-25, 27-30, 138, 182-184
Plate tectonics 29
Pollan, Michael 60
Population 15, 40, 41, 48, 96, 170, 224, 226, 227, 241, 252, 255, 257, 266, 267, 270, 275, 284
Positive culture 248, 266, 269-272
Postmodernism 151, 179
Pribram, Karl 197
Proto-cities 46, 162, 163
Proto-star 25
Protophenomena 65
Psychedelics 60
Purpose ii, 13, 14, 17, 133, 136, 145, 191, 209, 224, 229, 239, 244, 246, 248, 249, 265, 269, 279

Qualia 55, 57, 66
Quantum theory 17, 69-72, 74, 174, 175, 177, 194, 195
   summary 174
Quick Coherence Technique 218

Ramphele, Mamphela 242
Randall, Bob 193
Random walk 13, 15, 17
Raworth, Kate 228, 230, 231, 233, 236, 237, 243, 246, 298
Redshift 21
Reductionism 17, 18, 176, 211
Reincarnation 80
Restorative justice 255, 256
Richerson, Peter 31
Right purpose, definition 249
Rockstrom, Johann 1, 232, 236

Savage-Rumbaugh, Sue 118
Schrödinger, Erwin 69, 174
Schrödinger wave equation 71, 174
Self 54-56, 62, 64, 66-68, 77-80, 85, 86, 89, 90, 92, 99, 100, 103-106, 110, 112, 115, 124, 125, 190, 201-211, 213, 277, 279, 280
   Autobiographical self 64, 67, 104
   Core self 67
   Egoic self 110, 112, 203-211, 280
   in Buddhism 79-80
   Proto 64, 67
   -actualization 207
   -transcendence 208
Semantic Memory systems 119, 120, 124
Seven Markers iii, v, 187, 188, 268, Ch. 6
Sibbet, David 146, 147, 282, 298
Sixth Extinction 257
Skandhas 78
Social hierarchy 46, 49, 162, 163

Special Theory of Relativity 131, 173
Spencer, Herbert 11
Spirituality vi, 93, 131, 258, 260-262, 289
Stapp, Henry 70
Stars iv, 19, 20, 23-30, 182, 194, 264
   Distance to stars 24
   First Stars 23, 28
   Formation of stars 24
   System 25, 30
   What are stars? iv, 24
Steffen, Will 232
Stone tools 32-36, 48, 49, 61, 91, 118, 123, 140, 159, 162, 223, 274, 275
Strong nuclear force 25
Supernova 25, 26, 28
Sustainable Development Goals 232

Technological bottleneck v, 182, 185, 186
Teilhard de Chardin, Pierre 160, 187
Tenochtitlan, siege of 252
Tesla, Nikola 199
Thalamus 63, 64, 67, 85
The Ever-Present Origin iv, 54, 92, 94, 100, 111, 113, 114, 126, 212, 278
The great flash 23
The Great Leap 40, 42-45, 48, 91, 102, 105, 106, 124, 128
The Grid 141-144, 148, 199
The Reflexive Universe v, 131, 132, 136, 145, 279, 281
The Turn 133, 138, 139, 144, 145, 147, 148, 151, 282
Theory of Process v, 130-147, 149, 183, 185, 279, 280
   applications 146
   summary 147
Theory of relativity 131, 173, 281
Thompson, Evan 203
Thompson, William Irwin 97, 100

Tibetan Book of the Dead 80
Till, Emmett 254
Time freedom 112, 113, 198
Tononi, Giulio 57
Tree of Life 10
Trickle-down economics 230
Truth and Reconciliation Commission 256
Twenty-first Century Economics 238, 239

Ubuntu 193
Uncertainty 133, 136, 160, 174
Uncertainty principle 174
Unified field 198, 199, 220
Upanishads 75, 77, 165
Uruk 46
Ussher, James 13

Vajrayana 76
Vedas 76, 198
Verition 113, 114, 180, 280
Villanueva, Edgar 253
Voluntary memory retrieval 118, 124
Voluntary recall 121

Wallace, Alfred 10
Wave-particle duality 172, 174, 175
WEAll 243, 244, 247
Wellbeing Economy Alliance 243
White dwarf 25, 26
Wholeness and the Implicate Order 174, 176, 211, 212, 277
Wigner, Eugene 71
Wilber, Ken 92, 97, 129, 151, 166

Young Earth 13
Young, Arthur M. v, 3, 131

Zero-point energy 194, 195

Acknowledgements

I am grateful to many people who contributed to this book. My deepest thanks:

To critical readers Bill Briggs, Emily Silver, Stephen Jones, Chris Panym, Charlotte Sorenson, and MaryAnn Briggs whose suggestions helped me improve this book immeasurably.

To Emily Silver for her artistic contributions, and for always giving me a different perspective.

To Kendall Aleckson, Jim and Janice Hill, Cherie Arnold, David Hunt, and Bob Carmichael for encouraging me and believing in the project.

To Beau Rezendes for careful reading, many helpful suggestions, and unwavering support.

To Merlin Donald for kindly and patiently walking me through the nuances of his work and for giving me many of his published articles and book chapters.

To Joan Schleicher, Arthur Young's protégé and keeper of his works at Anodos Foundation, for constructive suggestions and warm support for this project; and to others from the Arthur Young community of scholars who were so generous and encouraging, including Jack Engstrom and Chris Panym from the Institute for the Study of Consciousness. And special thanks to Bob Whitehouse, who introduced me to Arthur Young.

To David Korten for his big-hearted help and generosity, and for his inspiring work.

To Chiaki Kobayashi for providing graphics and for answering my astrophysics questions.

To HeartMath Institute for providing graphics and for permission to use their work.

To David Sibbet and Kate Raworth for generously providing graphics.

To Sina Simantob who supported me in developing this book as a Scholar in Residence at Highland City Club.

To Kassandra White, Nicole Sturk, Crystal Edwards, and the team at Atlantic Publishing for believing in my manuscript and turning it into a book.

To the late Chip Chace for many discussions on consciousness late at night while driving out to the desert for climbing adventures, for giving me many of the most influential books I have ever read, and for keeping me honest.

To Jim Collins and Joanne Ernst for a lifetime of friendship, inspiration, and support that made this book possible.

About the Author

Photo: Bob Carmichael

Roger Briggs has spent a lifetime in the mountains of Colorado and from a young age had a love of nature and science. After studying physics and classics at the University of Colorado, he jumped into the world of public-school teaching at age 22. For the next thirty-two years he was a team builder, program developer, curriculum writer, and a beloved teacher wearing many hats. During this time, he established himself as one of the pioneering rock climbers of his generation. Since retiring from the public schools in 2005, he wrote the book *Journey to Civilization: The Science of How We Got Here*, a seven-year project, and he founded and grew a local non-profit for public lands stewardship. *Emerging World* is the culmination of his lifetime of seeking and research that brings both a cautionary and optimistic vision of the future of humanity. Roger lives in Boulder Canyon with his wife MaryAnn and two lovable dogs, Luna and Jasmine.

Made in the USA
Las Vegas, NV
06 April 2021